HOW TO
BUILD YOUR OWN
UNDERGROUND HOME

HOW TO
BUILD YOUR OWN
UNDERGROUND HOME
BY RAY G. SCOTT

TAB BOOKS
BLUE RIDGE SUMMIT, PA. 17214

FIRST EDITION

FIRST PRINTING—OCTOBER 1979
SECOND PRINTING—APRIL 1980
THIRD PRINTING—JUNE 1980

Copyright © 1979 by TAB BOOKS Inc.

Printed in the United States of America

Library of Congress Cataloging in Publication Data

Scott, Ray G.
    How to build your own underground home.

    "Tab#1172."
    Includes index.
    1. Earth sheltered house—Design and construction.
I. Title.
TH4819.E27S34   690'.8'6   79-16854
ISBN 0-8306-9744-6
ISBN 0-8306-1172-X pbk.

Preface

Prehistoric man became the first underground home dweller by taking refuge from the extreme elements in caves. Even though materials and methods have advanced drastically, I will be the first to admit that subterranian home building has not appealed to the masses of modern times. Its concept is just as logical today as it was back in time, but the catalysts today are financial and ecological because of the continually rising costs and increasing shortages of fuel and electricity. The following is my method of constructing a comfortable, modern geothermic home.

This book is not intended to provide all the technical information included in other publications, such as *Earth Sheltered Housing Design* prepared by the University of Minnesota. It will provide you, the reader, with an insight into the problems which a private home builder might encounter when tackling an unconventional project such as building an underground home.

Ray G. Scott

Acknowledgements

I want to thank my wife, Juanita, and my three children, Michele, Vicki and Chris, whose physical help and encouragement kept me going through another unusual phase of my life, and to my mother for her usual help. Also to Steve McCoy who was always around when I needed him and to the following good friends (In alphabetic order): Larry Bagwell, Curtis Barber, Lester Bull, Nick Crawford, Bill Ertel, Werner Ferrone, Bob Gentry, Don Greenfield, Gurvis Jones, John Neukam, Barry Scott, George Smith and Family, Roger Thompson, Stan Wales, and to these Commercial Businesses: Montgomery Ward: Bel Air, MD; Ralph Currie Carpets: Cockeysville, MD; Classic Kitchens: Bel Air, MD.

Contents

Definition and Objectives

The definition of an underground house at first seems totally self-explanatory, but if you consider the limitless variations of terrain, weather and human nature, you can imagine the extremes that are possible. Therefore, a clarification, along with my definition, is in order here at the beginning of this book. You can imagine the extremes that might occur when somewhere, sometime, someone has set up a permanent residence in a cavern or cave, 100 feet or more below grade surface. On the other hand, there are homes, particularly in California, with approximately 4 inches of soil or sod growing on the roof only to act as an insulator against extreme hot and cold temperatures. These two conditions should definitely be considered the extremes of engineering ease and difficulty.

Therefore, to avoid covering problems and methods of construction that would probably never be encountered by a potential underground home builder reading this book, I have established my parameters of a typical (if there is such a thing) underground home to be 2 to 5 feet under the earth's surface. The biggest reason I have for suggesting 5 feet of earth over the roof of an underground home is that this is the point of best compromise. By compromise, I mean that 4 to 5 feet of earth gives you the most insulation for the least amount of weight. If you consider that earth, with an average amount of small rocks and dampness from rain, weighs around 100 pounds per cubic foot, it is easy to calculate how many cubic feet of dirt you will have over your head. Multiply the number of cubic feet by 90 pounds. This will give you an estimated total weight that your concrete roof slab will have to support. The more weight overhead, the more reinforced concrete you will need and the more it will cost.

According to my personal tests and calculations, 5 feet of earth will give approximately 90 per cent of the insulation value that 10 feet of earth will give you. However, the cost and strength of the roof slab to hold up 10 feet of earth would be unreasonable, probably three or four times more expensive than a slab of concrete capable of holding 5 feet of earth. Figure 1-1 is a bar graph estimating the per cent of insulating value in relation to depth of soil.

Just as the depth into the ground could be varied, the amount of vertical exterior wall surfaces covered by dirt is likewise varied. The previously mentioned sod-roofed homes have no exterior walls covered by dirt. All the walls are conventionally constructed. But the home which I have built and now live in has three and one-half vertical sides covered by a minimum of 4 feet of earth. In all fairness, I've seen a few good designs with only three walls covered by earth. So now having mentioned the extremes for exposed exterior wall surfaces, I'll make this suggestion: If you are really interested in building underground, go all the way. The engineering problems are basically the same regardless of whether you have one side, two sides or all four sides plus the roof covered by earth. However, the benefits are substantially increased in direct proportion to the more exterior vertical wall surfaces covered.

MENTAL AND PHYSICAL PREPARATION

If this definition doesn't scare you, consider one more thing. Before you make the final decision to build under-

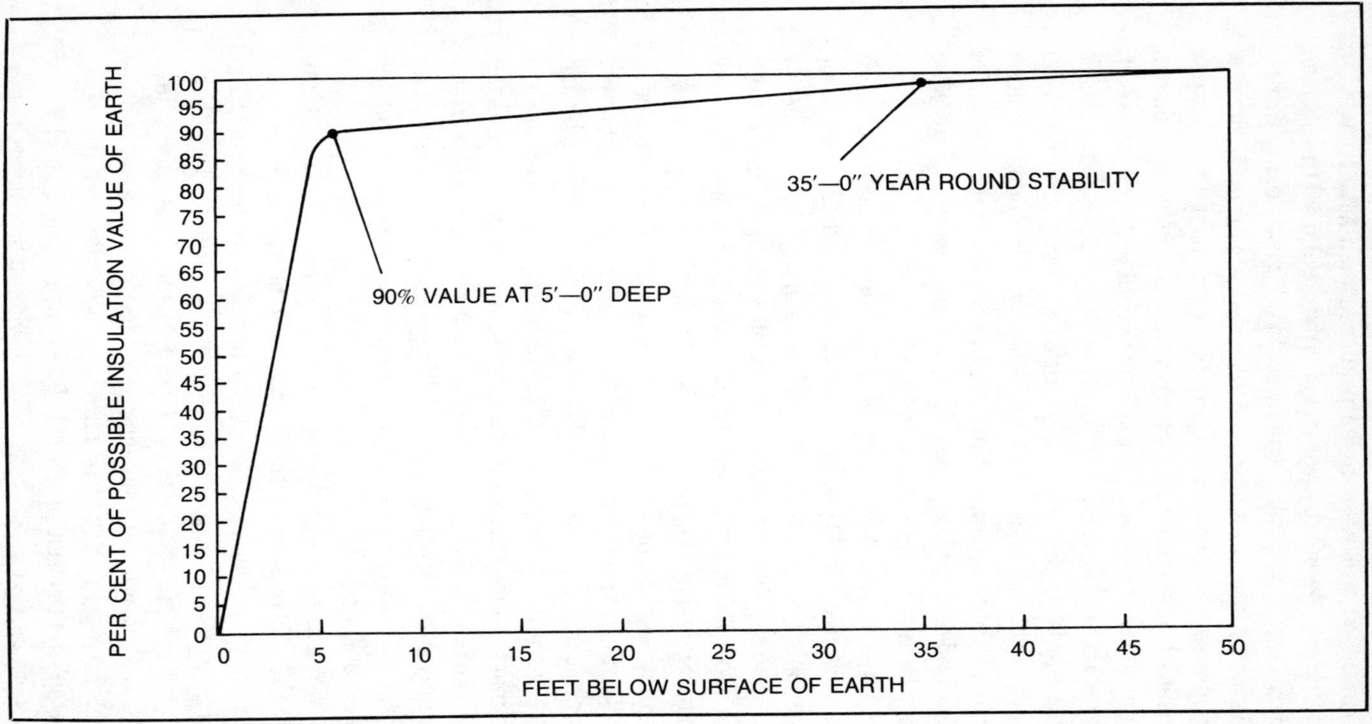

Fig. 1-1. Insulation value of the earth.

ground, you will want to be prepared mentally, as well as physically, to handle a project that is different from the normal. When I say physically and mentally, I mean exactly that. Building a house of any kind will test a person's physical stamina, especially when he tries to do most of the work himself. Remember that a concrete block soaked by rain can weigh nearly 100 pounds, that a shovel of wet dirt can weigh over 25 pounds, that a cubic foot of wet concrete can weigh 100 pounds a cubic foot, a sheet of plywood can weigh 80 pounds and that sheet rock might weigh 120 pounds. These are just a few examples of material weight. See the Appendix at the end of this book if you think building a house of any kind won't be a physical endeavor.

So now you can just imagine the kick-in-the-head when some stranger (or friend) comes along and verbally tears your plans apart with inaccurate and incomplete knowledge or facts, after you have been working all weekend, or a distant relative drops in after a hard day's work and proceeds to tell you that you have a screw loose and probably means it. This is an example of the mental harassment you must be prepared to handle. Sometimes people can unintentionally be downright demoralizing. But then there's always a friend to come by and lift your spirits by telling you that you are doing a good job and how much he likes the idea of building underground. This is the type of friend you need. He is also most likely the same friend you can call on for a helping hand. The person who knocks your project probably won't hang around long enough to get involved or help out.

Contrary to popular belief, most people are not innovators or experimenters, and they don't know how to handle anyone who attempts to be one, except by criticizing. This is probably the truest statement you'll read in this book. In my short life of 38 years, I have so far done quite a few unconventional things; building an underground home is definitely the biggest risk from a financial viewpoint, but it probably won't be my last project. If I've learned one thing by my experiences, it is simply that the overwhelming majority of the population only wants to take a look at the unconventional. Many people will say they would like to do this and that, or someday they'll do such and such, but they really can't come up with a reason for not starting their dream projects immediately. They just procrastinate until it's too late in life to accomplish anything, whether it be an underground house, a sailboat, a trip or a job change. As you read this paragraph, I'm sure you will recognize yourself, friends and family. I hope you, the reader, are daring enough to be innovative whether you decide to build an underground house or not.

I am fortunate that most of my neighbors were and are kind people and seem to be sincere in their friendship and interest. But don't count on this attitude of acceptance of your endeavor. Expect to be called foolish, dumb and worse. If you are lucky and you have nice neighbors, things should go well. Consider yourself fortunate. But I will bet you my last dollar that there will be one joker in your neighborhood, just as there was in mine.

BENEFITS

Now that you know what I consider an underground house to be, and you have been warned of the mental harassments (I'll tell you of more physical problems and legal pitfalls in subsequent chapters), let me now tell you of the benefits you can expect to find.

Fuel Savings

Since temperature, rainfall, winds and storms vary from coast to coast, some of your major objectives may be different from mine. The reason I decided to build my geothermic (sounds more technical than underground) home was definitely the cold winters and the hot summers of Maryland. In the winter the fuel bill in my previous conventional type home was doubling almost every year, and the electricity to run air conditioners in the summer wasn't doing any better.

Nature's Gifts

If you have ever taken a tour of any of the commercially-operated underground caverns, you'll remember seeing a sign somewhere near the beginning stating that the temperature at this point never varies from a specific degree, usually 54° or 55°F, winter or summer. This is one of the great gifts of nature which few of us take advantage of.

As you are probably aware, windmills and wind generators are enjoying a rejuvenated popularity since the fuel shortage of 1973. This is one way of taking advantage of nature's free gifts. Another, of course, is water power to turn a similar generating system. And don't forget that a few of the real back-to-earth folks (who deserve a great deal of credit) are using block ice frozen in the winter to keep cold storage areas cool through summer. As for solar power, I'll only mention it and suggest that you read up on the subject before building this underground house. There are millions of words written on solar energy. It's here now and forever, and it's practical to use.

These gifts of nature, along with many others, are used by only a small segment of the population because it's just not as convenient as they would like it to be. Geothermic heat and earth insulation are just as free, but only tested and used by a minute few. I feel, however, that this will be changing in the near future simply because the cost of all fuels will continue to climb at an unreasonable rate.

Once you are closed in 4 or 5 feet underground with average exterior exposure, you can expect to find year round temperatures stabilizing. Actually the temperature does not stabilize year-round until you are between 30 and 35 feet below the earth's surface, but the interesting fact is that after only 4 feet of dirt and 10 inches of concrete the lowest temperature I have recorded inside my home while under construction during the winter of 1977 was 46°F. This was without man-made heat of any kind. So instead of paying to heat your house in the winter from the mid teens to a comfortable temperature near 72°F, your additional heat requirement will be minimal. Remember, the constant temperature referred to underground is only when there is no life activity. Once you add light bulbs, cooking heat, body heat and appliance heat, the temperature will be much higher and thus leave only a few degrees to be raised by conventional or experimental means. Also note that in the summer these same heat additions are not great enough to require air conditioning. The highest temperature recorded in our home during the summer of 1978 was 79°F. You must realize that every building and every location is as different as the people who build them, so you can expect some of these examples to be more in your favor, or less in your favor depending on your situation. I am only quoting from my own personal experience.

Maintenance

During the time my working design was forming, I nearly overlooked the other major advantages. Consider exterior maintenance. Since an underground home has no trim to paint, windows to wash, or shingles to blow off, and the exposed walls are stone, you do not have to put many hours into constant outside work. The only thing you really need is a good riding lawn mower and a small push power mower.

Permanency

As for a third reason, don't forget that all exterior walls are concrete, as are the ceiling and floor. So if they are designed and constructed correctly, they are impervious to nearly everything. Nothing—insects, fire or water— deteriorates the basic structure. So you can forget the annual termite inspection, rotting beams, etc.

Equally important is the elimination of a major fire potential. If you build your house as I did mine, it is nearly impossible for a fire to get a foothold unless you are a pack rat and cram your storage areas with combustible items. Of course, even an underground home would be vulnerable to this type of carelessness.

Theft Factor

By eliminating all windows, you remove the temptation of vulnerable openings to a petty theft even though we know that the pro is going to find a way to steal regardless of the type of house you have. However, an underground house definitely gives you a feeling of security and stability.

For the Lady of the House

Remember when you don't have windows on every wall, you can arrange furniture in an endless combination because you never block a window. This should be particularly interesting to the woman of the household. Also, she will delight in the fact that she will not have to wash windows, nor will you have to buy as many curtains as in a conventional home.

Property Conservation

Last, but not least, is one of my favorite reasons. I bought approximately 2½ acres of rolling hillside in beautiful Harford County, Maryland. After building a 3600 square foot house, including the garage, I still have 99 per cent of the acreage usable and unobstructed by any man-made objects. Besides, the children can't knock a baseball through my windows.

Now that you know all the good points and are thinking seriously about a subterranean home, as they are sometimes called as opposed to geothermic or underground, I'll tell you a few of the alternatives you can begin to think about.

ALTERNATIVES

The alternatives are unlimited—just let your imagination take over. For example, these houses can be circular, rectangular or square. There are one-story, split-level and probably two-story houses. They are as small as one or two rooms, or as big as my 40 foot x 90 foot, two-level house. Size and shape are only the first of many major decisions you will have to make.

Skylights or domes are usually necessary, sometimes covering indoor gardens, sometimes only providing light to rooms. You could have one dome or six skylights or anything in between. One underground house I know of has the main entrance by a staircase to the center of a circular layout. I consider this an unconventional underground house because of the irregular room shapes. Room locations and exits are your personal choice. The method of construction is what is critical. The most important thing to consider is that room location will have to meet building codes where applicable to your locale. I'll cover codes in a later chapter—read it carefully.

Another factor that enters into alternatives regarding your house design is the land or site you select to build into. Does your design fit the land? Remember, you can build an underground home almost anywhere except a swamp. Once the exact location is established you eliminate many of your alternatives. For example, if the land you decide on is rocky, you will not be excavating as deep as you would if it were a sandy soil with no rocks. And if it faces north instead of south, you certainly would not want to build the open side of glass; just as you shouldn't excavate very deep if standing water is close by. One fact you will notice as your plans begin to progress is that by the time you meet building codes, zoning regulations, neighborhood restrictions and avoid the natural pitfalls, you don't have the limitless possibilities you first pictured. Nevertheless, even after meeting all these regulations, an underground house will still be a challenge to the imagination.

Keep tossing over in your mind all the things you could add to make your home interesting and individualistic.

2

Design and Land Must Work Together

The design of your house and the contour of the land must work together. This is true regardless of what type of home you are building, but never more true than when designing an underground house. It's a fact that any land with a water table low enough to build a conventional home could be used to build an underground home, but a water table varies from site to site. The water table is the depth underground at which you first come in contact with standing water under normal prevailing conditions. It is also a fact that the drier the land, the easier your underground endeavor. Whatever you do, don't buy the first piece of ground you come across. Look around until you have found a couple of acres you really like. If there is such a thing as the ideal location for an underground house, it would be the top of a knoll or hill in the high section of your locale. It is also obvious that you can't always get the best, but since this is one of the most important decisions in your life, make it the best possible. Look very carefully at the drainage adjacent to your potential house location. Pretend that it has been raining for a week and imagine where the run-off would be going. This is the first step to judging the usability of the land.

WELL DEPTHS

Next, check the depth of some of the wells drilled in recent years adjacent to your choice of property. You will be looking for places where well depths are 200 feet or deeper. This tells you that water is not near the surface of the land you will be building on. This is not to say that if the neighbors have a well only 100 feet deep that you can't build an underground home nearby—it's only an indication that moisture is closer to the surface, thus nearer your potential house.

EXCAVATING

The type of problem you could run into is this. If by chance, you started excavating in mid-summer when most sections of the country are entering their driest months, you could be deceived into thinking the land is really ideal, then when the following spring rains begin (as they always do), the land around could turn into a big swamp, holding water like a sponge. If this would happen, of course water would try to come through the concrete wall, floor or roof.

WATER PROBLEMS

The action of water coming through a concrete floor of any house is hydrostatic pressure. Briefly, this is when the soil around the concrete block cannot absorb any more water and the excess water cannot run off because it is surrounded by a less porous-type soil or rock formation. At this point, the water has nowhere to go but through the path of least resistance, usually your concrete floor or block wall. This is a condition you must avoid at all costs when locating your underground house. If possible, wait until you have a rainy period. The greater the rainfall, the better. A positive way to see what the conditions would be around your house if it were already built would be to dig a hole 20 to 25 feet deep. It should be big enough to climb down into, but be careful of caving walls. Examine the soil close—under actual conditions. If you don't have time to do all this investigation, or it is not feasible for some reason or you would prefer to have a professional give you advice, look in the yellow pages of the telephone book under engineers—soil testing or engineering—soil. Every community of a reasonable size will have one or more listed.

Anywhere there is major construction going on, especially road work, there will be soil testing facilities close by. This will cost you, but it is definitely worth the expense if the land you're thinking of buying is questionable. Remember, water leaking in your underground house could make it useless, and you would lose everything that you have invested.

COUNTY REGULATIONS

Last, but not least, your county probably has a department called *Land Use* or some department with a similar title. These local offices are usually a good source to check once you have limited your search for the land to one or two parcels. They have topography maps in great detail concerning the lay of the lands around your proposed home site. Also they have soil testing performed by Farm Bureaus for crop growth. All this information will be helpful to the person trying to make up his mind about a piece of ground. Also, the Health Department in most locales could be helpful, depending on their function within the local government. Be prepared to meet resistance or limited cooperation if these local officials know you are contemplating building underground. My suggestion to you, at this stage, would be to keep your plans to yourself. Remember, this is the voice of experience talking. Use all these sources to their fullest, combine them with good judgment and you shouldn't go wrong.

I strongly suggest that you make the final decision on exact location and get the deed in your name before beginning the design of the building. The reason I say this is that many things can happen on the way to the lawyer's office for final settlement.

FINANCING

You may not get financing at a reasonable rate, or you may not get financing at all. Let me tell you what happened to me because it may very well happen to you and you can plan accordingly. Since I'm building my house in the community that I was born and lived in all my life, as did my parents and relatives before me, you can see that the local banks were definitely on a first name basis with me. Aside from my life-long residence, my credit was flawless. When I decided to build a new house (my underground intentions as of yet unannounced), I stopped by to see my friendly banker. After a short discussion on houses in general, he asked me how much I needed. I gave him my figure and he said no problem. Within a few days a letter came in the mail stating that I could have the amount I requested—just stop in when I was ready to finalize the loan. It was just that easy, even though I was going to be my own contractor. This fact may bother some banks, especially if you don't have the credentials to back up intentions of being your own contractor. But this didn't bother my bank. Remember now, to this point they know nothing of an underground house. So I have the letter of loan approval, but as I am a basically honest individual, I decided to tell the bank of my plans to build underground. As a matter of fact, I even built a scale model of my house (Figs. 2-1 and 2-2) to impress the vice-president. He was impressed all right—so impressed that he said he couldn't possibly approve a loan for a far-out

Fig. 2-1. A scale model of the outside of an underground home.

Fig. 2-2. A scale model of the inside of an underground home.

venture like an underground house. At this point the negotiations began and after a great deal of pleading and promising my life away, they changed their minds and approved my loan. But this was only because of my life-long ties to this particular bank. If a stranger or a younger, less proven individual approached my bank or any other loaning institution for money to build underground, I'm afraid the answer would be short and sweet. No, No, No. This story is not meant to discourage you, but only to emphasize what you are up against mortgage-wise. As for V.H.A. and F.H.A. loans, they are even more difficult to obtain. If you qualify, check into them, but don't waste time begging or waiting for government assistance.

Another problem could be that the restrictions may not allow an underground house, or the soil may be unsuitable. These are just a few of the potential roadblocks to actually getting a deed in your name. If you invest your valuable time and money designing a building for a particular site and the deal falls through, it would be unfortunate and costly. If a professional draws up your blueprints, he will adapt to a particular piece of ground. If you change locations, the prints would have

25

to be revised to suit the new location. This is not necessarily true for a conventional house, but most likely for an underground house.

LOT SIZE

I also suggest that the size of your lot be no smaller than two acres. The reason for this statement is purely cosmetic. Conventional homes can be designed to be attractive side by side on small plots of ground, such as many developments are,

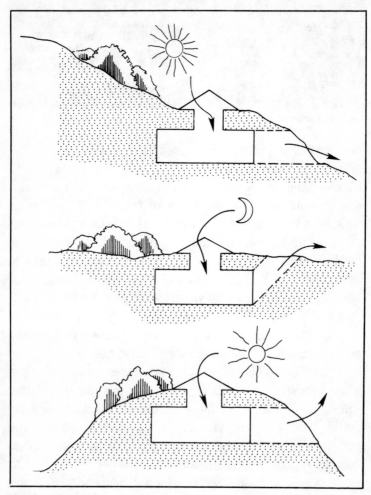

Fig. 2-3. Three basic possibilities for location of an underground home.

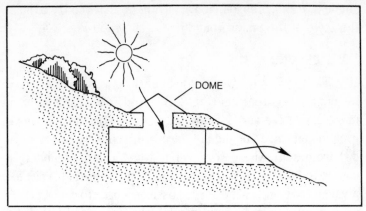

Fig. 2-4. Into-the-hill method of building an underground home.

but in my opinion, an underground home loses much of its appeal when crowded by conventional homes.

Now let's assume that your dream location is a reality. The deed is in your name. If you have picked out a good piece of land, 50 per cent of your potential problems will be eliminated. See Fig. 2-3 and locate the lay of the land that most resembles yours. Note the house location in each situation as I will explain the pros and cons of each as I see it.

INTO-THE-HILL

Into-the-hill is the most popular approach to building underground (Fig. 2-4). This into-the-hill method is by far the easiest to build, especially from a grading viewpoint, because approximately 65 per cent of the cubic footage has to be excavated. The reason this is ideal is that you will need tons and tons of dirt to backfill and complete final grading. This into-the-hill method provides a natural method of moving building materials around by an ordinary truck, because the upper part of the hill allows you to work at roof level. At the same time, the natural slope will give you access to the lower level by normal vehicles. If it were not for the ability to deliver building material to the lower level by truck, you would have to keep a crane of some type on the site to lower the heavier building material, and those cranes are expensive to rent. I suggest, for this reason, above all others, that you try to build the style of the underground house shown in Fig. 2-4.

LEVEL GROUND

The *level-ground method* is to be used as a second choice for quite a few reasons (Fig. 2-5). First, backfilling is much more expensive and difficult because the structure is above ground and since you did not excavate, you have to obtain fill dirt from somewhere to cover it with. This is definitely a course to take as a last resort. I had to buy some extra dirt for my final grading. I called every possible source to check prices of a dumptruck load (approximately 14 tons) and I found the prices (in 1978) ranged from $3 per ton to $10 per ton for the best quality top soil. So you see, buying the good earth isn't a cheap approach. Even if you could get dirt for free, you would have to rent a crane and bucket to put the dirt on top, because it would be unwise to take a bulldozer on your top slab, regardless of its designed strength. This would be expensive and time consuming. The extra equipment combined with the cost of buying and hauling soil will make the style more expensive than you probably want to get involved with. A second point to consider when looking at the ground-level style is that the heat loss and heat absorption is greater when a mass of earth is less in volume and above the natural terrain. There is

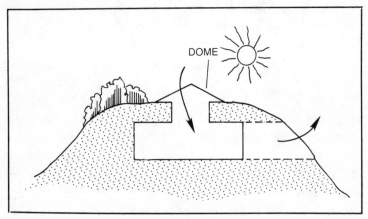

Fig. 2-5. The level-ground approach to an underground home.

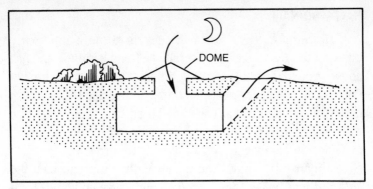

Fig. 2-6. Below-grade level method of construction for an underground home.

one positive aspect of this type. It may be easier to comply with local building codes, simply because you would have the ability to exit from each exterior wall with limited trouble and cost. Also you eliminate the cost of the initial major excavation.

BELOW GRADE LEVEL

The *below-grade-level* is, in my opinion, the least desirable approach (Fig. 2-6). The major reason is that exit and entrance would have to be up and down a set of steps, and a garage entrance would be at the bottom of a hill. This presents a drainage problem, unless a well-designed drain system is installed. In many cases an auxiliary sump pump would be required to pump storm drains up to a natural drain level. In almost any situation where the house is built below a level grade, the drainage system becomes a major endeavor and also very expensive. Consider the possibility of a clogged drain or power blackout. Your house would be flooded. Not a nice thought, but possible. In addition, the pump on this system will require occasional maintenance. If these features don't bother you, then go ahead and use this method.

The construction is basically the same in methods one and two. If there is one benefit to the below-grade-level house, it is that you will not have to haul in additional fill dirt when you are doing the final grading.

HALF-AND-HALF

Half-and-half is a term I've given to the many homes built, as the title indicates, literally half underground and half above ground (Fig. 2-7). At first thought, you could argue that there is no difference between these homes and thousands of conventional homes that have used their club basements as a daily living center. I guess this is technically true, but if you look closely, you will have to agree that there is a real difference. Figure 2-7 shows the major feature that qualifies the *half-and-half* house to be called an underground home. Usually the structure above grade level is used for a garage, storage, an area for solar heat storage tanks or a possibly a work shop. In all fairness, the half-and-half homes researched for this book had less than 20 per cent of the actual floor space above ground, and that 20 per cent was always a nonliving area. One of the advantages of this style is that exits and entrances are more conventional, thus easier to meet building codes on local restrictions. If there are disadvantages, it would be the extra cost of working on two levels and extra precaution that would have to be taken to insure against water seeping along the first level into the lower level.

BASIC NECESSITIES

Now that you're getting down to serious business with this geothermic home idea, don't overlook the forest for the

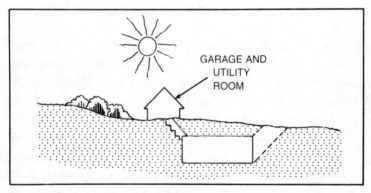

Fig. 2-7. A half-and-half underground home.

trees. You're expecting those unusual problems to show up when you do something as different as this, but don't forget the basic necessities that should be considered regardless of the type of house you are building.

The accessibility to good roads in winter or rainy season should be considered seriously. It's easy to say you like remoteness and getting back to the wilderness, especially when the weather is good and you're excited about a new project and are physically active, but this house will be there a long time and so will the mortgage payments. Stop and think how you will feel about a specific location in 10 years. You've settled down a little, probably have a few more grey hairs and the children are growing like weeds. That beautiful, picturesque, four-acre tract of land you got real—cheap 10 years ago because it was 5 miles to the nearest paved road now presents the problems of daily trips to the store and children walking to meet the bus. This may not be the ideal location you first thought it was.

Also check into the cost of running telephone cables from the road to your house. The cost varies from location to location, as does the method by which the electric company charges you to bring power to your house. The way it works in Harford County, Maryland is like this: the Electric Company will install service lines to the meter, regardless of the distance from the existing power source at no charge to the home owner, providing the meter is installed on the nearest possible corner of the house. If you want the meter on a wall or post farther away than the nearest point, you have to pay the Electric Company so much per additional foot of power line required. Ask precisely what the charges will be. You will find the telephone and electric companies very cooperative organizations when planning the location of lines, meters and the place of line entry into your underground house. I won't go into gas lines because I emphatically suggest you don't even think about liquid gas, propane or natural gas as a fuel. As you know, a leak in a gas line is a serious problem anywhere, but in an underground house, it's more of a concern because of the air tightness of the structure. Don't ask for trouble with these fuels—building underground will present more than enough problems. If you stick strictly to electricity as the utility and a wood or coal burning stove for heating assistance, you'll be safer and happier.

Before signing your life away for a parcel of land, make sure water is available. Most likely you will drill a well, probably deep if you have a good piece of ground. One of the regulations in Harford County, Maryland is that an approved well usually must be drilled before construction can begin. You would be in sad shape to get your money invested in construction and find out there is no water down below. It's a necessity of life.

Well drillers are usually full of valuable information and they can probably predict the depth, quality and quantity of water you will find fairly accurate. Experience has taught them a great deal. So have faith in your well driller. You have the final say as to the location of your well, that is after the health department suggests a particular area. Make sure the well driller agrees with your choice. One important phase of drilling a well is to discuss before hand with the driller the method of payment. Most drillers have a set fee per foot for soft dirt and another for hitting rock, or they will quote you an average of the two prices, regardless of what they strike on their way to water. Also ask them about drilling a second or third hole, if the first or second turn out to be dry. Some drillers do the second drilling for half price. Some have other arrangements. Find out before you start drilling, not after you hit rock.

Once you get water in, you have to get rid of the waste it creates. So sewage disposal is just as important as finding water. Once again the local regulations dictate your septic system's location, size and configuration. Make sure your land is approved for a satisfactory system before you build, even if it is not a standard requirement. Ask for a perc test before buying a piece of ground for a private home, if public sewers aren't close enough to hook into. If the soil is not suitable for a good sewer system, it most likely is not going to give good drainage for your house.

Don't buy property simply because it's cheap. If you do, it will come back to haunt you. Sooner or later you'll pay for buying a marginally acceptable piece of ground. Remember, you don't get something for nothing. Last, but not least, check out all codes, regulations, ordinances and zoning laws that may apply to your choice of property to see if they will cause you a problem.

Some land has restrictions attached to the deed from previous owners. For example, years ago a dedicated farmer could have stipulated in his will that his farm land never be used for anything but farming. This can be done legally, and all of a sudden you could find yourself raising a herd of cattle instead of building an underground home.

Another source of restrictions comes from development preferences. If a housing development has a set of restrictions drawn up, they may include the size of the house, material used and style. I have never heard of one that allowed underground homes, so watch out. Zoning law will be the easiest to comply with. They simply control things like multi-family dwel-

Fig. 2-8. Be sure your land is zoned for a single-family underground house.

ling, commercial ventures or farming. It's very easy to find out if a particular piece of land is zoned for a single-family underground house (Fig. 2-8). Just stop by the county zoning department and look at their zoning map. I think every county in the nation has one.

Now if you are not discouraged, continue reading into chapter three. Remember, I'm trying to save you aggravation and trouble, not discourage you.

3

How To Get Work Done

Since you are building this house yourself, you certainly don't lack energy or interest, so there are only a few things that can slow you down. These include the weather, lack of money or poor planning. The first you have no control over. Money is a personal thing between you and your bank. You know how much you have to work with and how careful you must be with it. This only leaves planning. This you have total control over. Whatever you do, don't confuse planning with designing. Designing is physically locating material items. Planning is the art of allocating your time.

Design will be settled long before you actually begin to build, and you really can't do much changing after you have actually begun pouring concrete and laying block. Once your plans are approved by all necessary authorities and a building permit issued, you can consider the design phase of your project final, except in minor instances.

However, time allocation will continue from day one of construction to the time you use the last paint brush. So you see, this is the important part of building that you have control over. Time is as important as money. Sometimes it is more important because money can't buy time, but in time you can make money.

Once you're committed to building this underground house, figure on one year of continuous work on your part, maybe even more, depending on your desire to get the job done and nature's cooperation.

First of all, don't get behind before you start. Break ground in the early spring. If you wait until mid-summer, you will be in a race with mother nature to beat cold weather. Of course, I'm assuming that you're building the underground house in a region with extreme seasons, such as in Maryland. If you happen to be in the deep South or Southwest, you just have to appreciate our problems of changing seasons. The reason it is imperative to complete all concrete work and get it covered with earth is simple. Concrete expands and contracts like anything else with heat and cold. The expansion and contraction cause cracking which is a condition you need to avoid at any cost. This is why I suggest you plan carefully to be able to pour your roof slab during a time period where the temperature doesn't get below freezing or vary to extremes of hot and cold. You must realize that if your building were being built in a locale where the temperature might reach 100°F during a day in September, but fall to 40°F at night causing a 60°F differential, cracking could develop. This condition would probably do more damage than actually pouring concrete at 50°F and having the temperature drop to 32°F at night. The extreme differential is what causes the cracking problem and should be avoided at any cost. How's that for a planning problem?

I can't begin to tell you how to plan well personally. I don't think any book can. You are born with that ability. It's almost a talent—like singing. Fortunately, I was born with a planning ability a little above average. However, I know that if I had been even better at planning, my job would have been much easier. Just think a little bit ahead. Don't overlook little things, like small tools, nails or a water cooler.

A good example is something that happened to me. I had a gas generator on the site to use while building the scaffolding for the roof. I rushed to the site early in the morning with my power saw, started the generator and, lo and behold, I'd forgotten my extension cord. For the want of an extension cord hours of work were lost until I left the site and picked one up. Now I was ready to cut boards, and I did, for about 15 minutes. Then the generator ran out of gas. As if this wasn't bad enough, I didn't have a container to get gas in. So by the time I got a can out to the gas station, back to the site, started the generator and was ready to work, I was starved and ready for lunch. Of course, I didn't brown bag a lunch at first because I didn't realize how much time it takes to drive three miles, get a sandwich and soda and get back to work. I soon learned to brown bag a sandwich on those days in which I planned to put in a full day's work. If there is food and drink on the site, I found I could get an average of two more hours a day of working time. This really adds up in a year.

I have to admit that this day I described was unusual for me, but it is an example of what good planning and a little thought can prevent.

As for how to get additional working time, this is the eternal problem of a do-it-yourselfer. I presume the reader of this book to be working a regular 40-hour week to pay for the other 128 hours and the house. Let's also assume that you use 10 hours a week travelling to and from work and to and from the building site; you probably need seven hours of sleep a night; and time to eat. Let's say that is 10 hours a week. This gives you a total of 109 hours just to eat, sleep and work at your regular occupation. That leaves 59 hours to take care of personal business, see your family, socialize and build a house. So you see, there's really no time to waste in idle talk. You cannot waste effort of any kind.

One of the fine lines you will have to walk when building an underground house is tactfully dealing with and handling friendly, inquisitive people. This sounds like an unfriendly gesture, but it is not. Here's what I mean. Naturally an underground house is interesting. Maybe some day there will be enough underground homes around so that the majority of the public will have seen one or possibly have been inside of one. Once this happens, of course, the novelty will be missing and the curiosity seekers will not visit you, but this is definitely in the future era. Be prepared for the present day, because as the word of your innovative project spreads, your friends and neighbors will stop by to talk with you about your house. This is the problem. How do you politely keep on working while they are asking you basic questions? Remember, after a while, you have answered every question many times and explained the details to many people. But each time the questions are asked, it is the first time for your friend. Soon you will find yourself talking more than working. However, it is good to know that people are interested, so you just do your best to work and talk, talk and work. Most people will volunteer a helping hand and that is something you can always use.

INSURANCE

This brings me to another subject that should be mentioned in conjunction with an underground house. Insurance. With all the visitors you will have on your property from beginning to end, I suggest you obtain a good insurance policy covering personal liability. As I'm sure you are aware, it's awfully easy to have an injury on a construction site, especially in the early stages when the terrain is full of ditches, rubble, nails, loose timber and so on. Ask your lawyer about posting the property as a means of liability protection, but don't count on a *no trespassing sign* to protect you against a lawsuit if someone would get hurt, because it would be voided by your personal invitation to visitors or workers. You are still liable for these people. Anyway you slice it, you need a good insurance policy.

BEST WORKING HOURS

As for working time, I found the best time to work uninterrupted was to start at sunrise. I don't mean just early—I mean exactly at sun-up. Somewhere around 5 a.m. Once you get used to these early hours you'll find work progresses must faster than evening work. It's also invigorating and the sunrises seem to be good for the mind. To be honest with you, I found the weekends, holidays and vacation days to be the time to get the big jobs done. You'll find that the larger phases of construction need to be completed by continued hours of working. You can literally never get a big job done (for example, plumbing) if you do it 15 minutes at a time, one day at a time. It will take 15 minutes to get the necessary tools and supplies together and ready for work, whether that work period is to be for 10 minutes or 10 hours. This is another way to save valuable time by planning. So if you have only a few minutes to work, do something that can be completed in that time frame, if at all possible. One such job would be hanging a door. Or use the time to prepare a particular area for a bigger job. Keep this train of thought in mind and you will find you have saved days by the time you're finished building.

The early mornings and evenings I used to take care of small things. If you can arrange to take a vacation at times when a big job needs to be completed, then you are fortunate. There's nothing worse than having a complete week off from your normal job, have it pass and accomplish nothing significant. It can really demoralize and frustrate you if it happens.

Still another phase of getting work done is affected by availability of material. This is where planning far ahead on your part can prevent a problem. When ordering material, be sure to give the supplier long enough advance notice to deliver supplies by the time you need them. Don't wait until the last minute and then waste time because you are missing one 2 x 4 or one concrete block. Don't forget, you're not the only one ordering supplies. Don't expect the vender to jump at your request for immediate delivery.

If there is one way above all others to save time, it is to complete one craft throughout the house. For example, if you are putting furring strips on the block walls, start in one room and continue throughout the house. The same goes for plumbing, electric, sheet rock, painting or whatever. If you try to complete one room at a time, from scratch to final trim, and then move on to the next room, you will spend most of your time putting tools away and getting new ones out. Also by completing one job throughout the house, you become skilled at the particular job. The only thing to be said against this method is that it gets boring. For example, hammering nails for three days and then painting for three days becomes very monotonous. Jumping around can break this monotony, but it wastes valuable time.

Probably the most effective time saving idea that I will mention is following: When you have help available, whether it's a friend, relative, paid worker, young or old, male or female, don't overlook the fact that everyone can do some type of work. But, if you ask the wrong person to do the wrong job, you either have to hold their hand or do the job over. Don't ask a child to do a man's work and vice versa. This way neither person will become frustrated by being asked to do a job they can't handle. Use your labor, whoever or whenever, effectively.

About the only other thing which is really important as far as saving time and making work easier is to have work that needs to be completed inside and outside even after you are under roof. In Harford County, Maryland, just as in most other locales, the weather varies considerably. Even in mid-winter, we have occasional time periods of warmer than usual temperature. When this happens, get something done outside that can't be completed in bad weather, because in the warm weather there are surely going to be days that rain prevents outside work. I know of people who would just cancel that day's work. But if you're really anxious to complete your house, you can always find something constructive to do whether it's winter or summer, night or day.

4

When to Hire a Professional and Save Money

It is easy for me to tell you to plan ahead and save money when this was something I found most difficult to do successfully. Since you are reading this book, I have to assume that if you build, you are planning to do some of the work yourself. Should you decide not to build underground, this chapter will be a help regardless of the style of house you decide on, as long as you do 50 per cent of the buying and building yourself.

As we all know, there is always more than one way to do anything. This is never more true than when building a house. The crafts which I have found easy, you may not and vice versa. So as you read on, keep this in mind and pick the subjects you feel comfortable with.

PROPERTY LAYOUT

One of the easiest ways to begin your project and save a few dollars right from the start is to locate and lay out your house corners on your property, once you own it. All you need is approximately 30 wooden stakes about 1″ × 3″ × 48″, roughly 400′ of surveyor's string, a surveyor's transit, a plumb, a bob and a 100 foot steel tape measure. Most likely, a local rental company will rent a transit to you and can show you how to use it. There is no need to buy a transit. They are expensive, and you will only need this instrument in the beginning phase of building lay out. It is one of the few items that I suggest you rent as a method of saving money. In addition, transits around construction sites have been known to grow legs and walk away from their owners, never to be seen again.

After a short lesson with your transit and a few practice readings, you can lay out the corners of your house as well as anyone. Just go slow and double check yourself. Be sure to check local regulations regarding side clearances to your property line. Your house usually has to be at least 10 feet from your side boundary, but check the code to be sure.

By telling you to locate your house corners I am not suggesting that you try to locate permanent boundary survey markers. These are two different jobs, and this is a job for a professional surveyor. Also, most localities require a certified surveyor to check these boundaries and a practicing lawyer to record the deed.

LAWYER'S FEES

Lawyer's fees connected with buying property or recording mortgages are occasionally a rip-off. Ask friends who have recently purchased land or a house what their charges were.

One of the biggest surprises connected with home building or buying that everyone encounters is that—the actual cost of the property that is asked by the seller is only the top of the iceburg as far as cost goes. If anyone you know has recently purchased real estate, they will attest to this. I'm relaying this warning especially to the younger person buying his first property. Here is a totally mythical example only to show the progression of dollars required to buy a parcel of land to build on. Remember the figures and items are estimated because they vary drastically from locale to locale.

Sample Purchase

Here is an example ad in a newspaper: Two-acre lot; ½ wooded; nice view, $20,00. The sum advertised may seem reasonable to you, but don't relax and write a check yet. At the very beginning a perc test is required to insure the health department that a septic system will work well on this level. If this is required, add $100 to $200 depending on how easily accessible the land is. If you actually agree to buy the land, it must be surveyed. Add $200 to $500 depending on how complex the shape is. Next you need a lawyer for the title search. This covers hidden ownerships or liens placed against this property by previous owners. Add about $500 for a normal search. Of course, there is a tax just as in buying a car. In Maryland, there is a 5 per cent sales tax. So a piece of land costing $20,000 would be taxed for $1,000. Yours may be higher or lower. Now you have a registering fee. This is for recording your ownership on the courthouse records and everywhere else the law requires. This could run over $200 depending on where you are. Then there is a stamp tax. This is exactly what it implies. A little seal is attached to the documents as they go through the legal channels. This is a local tax as opposed to the state sales tax. Add $50.

Real Estate Tax

Of course, we must not be prejudiced against real estate tax. You pay real estate tax proportional to the fraction of the tax year that you will own this new property. For example, if the tax bill is $500 for a year, but you buy this property exactly halfway through the year, you owe one-half of that year's tax bill. Add $250.

Already this $20,000 piece of ground could be as high as $22,700, and you could continue to add little hidden costs that you might encounter. When I say hidden costs, I'm not indicating that the system is trying to deceive you. I only mean that these costs are not spelled out in big black letters to a first-time property buyer, and since most people don't buy homes and property on a frequent basis, I'm sure many people tend to forget what was involved in buying their property. Also the cost changes from year to year.

Also, these costs do not include the hook-up of utility, surcharges, highway access costs, permits or anything other than getting the deed in your name. These additional subjects will be covered in later chapters. Fortunately, I have a reliable friend who is a lawyer and another who is a banker, so I felt comfortable. However, I can't caution you enough to check for undisclosed costs. As for tax stamps, local taxes, fees, etc., you're stuck. They're everywhere and inescapable. So for now, accept them.

EXCAVATING

Now that you have your land, are happy with it and are ready to break ground, let a local excavator with a good bulldozer do the heavy digging and grading. Even if you could afford a dozer, you'd probably waste days learning how to use one well enough to dig a foundation. A professional can usually do a foundation in one or two days. Once the foundation excavation is complete, you can begin to do work yourself.

If you're dexterious enough to operate a backhoe, you can invest some of your money for a good return. For clarification, a backhoe is a rubber-tired tractor with a scooping bucket on the rear and a bigger scooping bucket on the front. Both are hydraulically operated and easy to operate with a little help from an experienced operator. These mechanical work horses are invaluable around a construction site. For example, after the dozer digs the hole and cuts a driveway, you can dig your own footers, spread your own gravel base in the driveway and move or drag anything that needs moving. These used backhoes are available (as of 1978) through dealers or private sales at a cost of between $3,000 and $8,000. It really makes no difference what the price is, as long as it's a good value. The point is that if you service it regularly and don't misuse it, you can get your full price back when you sell it after your home is complete. Remember, this same tractor will most likely do all of your fine grading, leaving only hand manicuring to complete your lawn.

The most important thing is that these vehicles can work close to your underground home and in many cases can work and grade even on the roof. However, do this only if your advising engineer says it is safe to do so. The uses you will find for this tractor are unlimited. Remember, shop carefully to find a good buy. Then don't panic to sell this vehicle immediately when you are finished because they always retain their resale value. You may even pick up a few extra dollars doing small digging jobs in the neighborhood.

BUILDING MATERIALS

Now that you've rented a transit and bought a backhoe, you're ready to buy building material. By building an underground home using my method, poured concrete and concrete block are the two biggest single purchases you'll make, so check supplier prices carefully. For example, my house required nearly 8,000 concrete blocks. When checking prices with the four major suppliers in my area, I found delivered price of a standard 12″ concrete block to vary as much as 15 cents a block. It doesn't take a genius to figure the savings here. If you have a method of hauling and unloading these blocks, you can save a bundle more, but most people don't have access to the equipment to move pallets of block (you need a heavy-duty forklift), so leave the job of delivery to the block compnay. Don't make plans to lay these blocks until you actually see them on your site. Block companies are famous for missing their delivery date by days, especially to a first-time or private home builder. The reason is simple. They have good intentions when the dispatcher says he will deliver on a particular day and time, but he knows you are an individual. Then a big contractor who is a regular customer calls an hour after you do and wants 5,000 blocks delivered the next day. You can easily figure out who is going to get their blocks on schedule. It's not you. By the way, most block companies have seconds. These are blocks with cracks or chips. They are just as good as the first grade once mortar is applied. Ask about them. They could be used for interior walls at an additional savings. However, don't use them on the exterior walls.

CONCRETE

Now for the concrete—remember you have to pour footers before you can lay block. (Refer to chapter 8 for details on how to prepare footers.) Once your footers are ready for concrete, don't even consider mixing you own. Buy from a local concrete delivery company. There are two standard methods that concrete companies use to deliver concrete to your site.

Cement Mixer

Once way is the conventional barrel-on-the-back type that you commonly know as a cement mixer. This type mixes cement, gravel, sand and water as they drive down the road or on your site. The only problem is that they are batch mixed. This means that if you order 6 yards of concrete, they load the cylinder with the appropriate amount of cement, sand and gravel. The only thing left out is the water. This is added when it reaches your site. However, the catch is that you now own and pay for all six cubic yards of concrete even if you over estimated and can only use four cubic yards.

Rectangular Tank

The second method of delivering concrete is the newer, rectangular-tank type. This type of concrete truck carries all of the ingredients—cement, sand, gravel and water—in separate containers. Once ready to pour, they begin to blend these components together forming a slurry mix of concrete only seconds before it comes out of the truck. With this method you only pay for what you use. Since everything is unmixed until immediately prior to pouring, the truck can stop pouring at anytime you designate and return to its home plant without wasting anything.

They do, however, charge a fee if you order a small amount, usually an additional 10 per cent. You will have to pay this fee if the total yards used from one truck is less than about 5 cubic yards. This second method also avoids any panic rush to pour if a problem arises, whereas the first method requires dumping to begin immediately after the water is added and continued pouring until the truck is completely empty, even if the excess is dumped in your driveway.

Since most companies in a given area will charge the same per cubic yard, the only other advantage you gain is service. If you are pouring your footers yourself, try to be ready for the first load of the morning from the concrete company. The reasoning is that this is the only load that will be on a known time schedule. After delivery of the first load, the driver returns to the company and loads up to begin his next assigned delivery. So the later in the day, the more delays he has encountered on previous assignments. Therefore, you could wait for hours for a load to be delivered. In all fairness to the concrete companies, it is difficult to stay on a schedule. Once I had a crew of friends over to help and the truck was three hours late, so we just stood around wasting time.

Again, if you have a choice, definitely buy from a company that mixes concrete on the site because you'll only pay for what you use. This is better than a pre-mix company because you order in advance and have to pay for the full load whether you use it or not. Contrary to what you estimated your concrete usage would be, you'll probably miss by at least a half yard due to inexperience and possibly an uneven pouring surface.

MASONRY

Now that your footers are poured and blocks delivered, let an experienced block mason lay your block. However, here's your chance to save again. Get a price to lay the block from at least three different small block laying companies with good reputations. Check their work and reliability in advance and don't pay for anything until the job is done to your satisfaction. It should take about three successive days of good working weather to finish the exterior walls of a standard underground house, if there is such a thing. Anytime you're working with a small contractor, ask his price. Once the work is done and you're ready to pay the bill, ask if he gives a discount for paying in cash. Sometimes they will give you a good discount for paying in cash, because they avoid the risk of a bad check. Also when dealing with a small contractor, get a clear understanding as to who is ordering and paying for sand, mortar and water. The total price is usually the same whether you take care of these items or the block layer does. It's just a serious delay to have block and workers, but no sand. If you can work with your block layer and have him agree to lay one day's worth of block around the total perimeter of you're house, maybe three, four or five block high (Figs. 4-1 and 4-2), and then leave the site and come back at a later date at no extra charge, it will be much easier to pour the concrete floor. If your mason has to continue laying block until the walls are at their maxiumum height, it only makes it harder to move men and equipment around to pour the concrete floor.

FLOOR SLAB

Back to saving money. Hire a concrete finisher to pour the smooth floor slab. This is one of the most important jobs in building your underground house. If the floor sets up uneven

Fig. 4-1. If possible, lay three courses of block, then pour the floor.

Fig. 4-2. Pouring a concrete floor.

or rough or cracked, it's a major problem because you will most likely be putting carpet and tile directly over the concrete and a smooth level surface is absolutely necessary. Find out who the contractors prefer. Right here I should mention that a good way to find the best people in the trades of plumbing, block laying, concrete finishing, etc., is to look for the one that can't give you immediate service. He's naturally in demand. The guy who is sitting by his phone and can start tomorrow is not busy for a reason. Find out why. I found it was better to wait for the busy man than hire an idle contractor. I'm not telling you not to give a new man a chance, but be careful. The concrete finishers will give you a price in cents per square foot. Once again, find out what this covers. How smooth should the pouring bed be? What about level pegs?

For clarification, level pegs are only wooden or steel pegs driven in the ground so that the top of each peg is level with the others and to the grade level that you want your finished concrete to be. As your concrete flows and you work it smooth, it should end up level with the tops of these pegs. This way you don't have to use any type of leveling device to insure a level floor or footer. If the pegs were put in accurately, then the floor should be level.

Pouring a slab of concrete is another way to save a few dollars. If your area is unlevel, fill low spots with gravel. Gravel is much cheaper than filling these holes with concrete. For example, a cubic yard of concrete, as of 1978, was approximately $35 delivered. A cubic yard of small stone would be approximately $3 delivered. This is quite a difference in cost if you're only filling a hole.

Fig. 4-3. Inside section of marked block knocked out to tie the interior wall.

INTERIOR WALLS

A second reason for pouring the floor after a couple of courses of exterior wall block are laid and before the interior walls are started is because of the difficulty of leveling and smoothing concrete one room at a time. There is absolutely nothing to be gained by pouring concrete after interior walls are up. Don't even consider this method.

A third reason to put interior walls up after floor slab is complete is that it is very easy to lay out the walls in their actual position. This is how I located my interior walls (Fig. 4-3).

Once your floor is finished, take the width of your interior wall block and cut a board that width. Usually in an underground home you can use 6″ concrete block for interior walls, unless your engineer says otherwise. This board should be approximately 8′ long for ease of handling. Now with a 100 foot tape measure and a can of bright spray paint (why not red) and a chalk line begin to lay the entire floor plan on the concrete floor. Spray along both edges of the board leaving a wall location mark (Figs. 4-4 and 4-5). This system is good because it gives you a chance to check actual sizes of rooms rather than looking at a blueprint. Plus, by this time you need a psychological lift and you will feel a sense of accomplishment seeing your rooms in full scale. Also, now is the time to make a change in wall location, not later. So walk through the make-believe rooms. Check out door locations, openings, closets, room sizes and anything else you can think of.

MORE NECESSARY PURCHASES

Now that you have things started, there's no turning back. There are a few more things to buy, preferably used. Watch for a bargain on an acetylene cutting torch outfit. They are easy to use and you'll need it to cut the metal rebar that you'll be putting in the concrete roof and retaining walls. They are hard to find, but you will need one and renting it is expensive. Plus, if you rent by the day, you are rushed to get

Fig. 4-4. Lay out interior walls with paint.

your cutting done and return the torch outfit. This is a good way to have an accident or make a stupid mistake.

Also buy a good radial-arm saw new or used. It will be one of the most valuable tools you acquire, and you'll probably keep it forever. Check prices closely and watch motor and blade sizes. Ask your friends and experts about sizes and

Fig. 4-5. Look at the actual room sizes as outlined by the paint. Make any changes in wall locations now.

brand names. Remember you're not a professional carpenter so don't buy the super contractor's model. Just buy the size you need.

Another thing I bought that has proved to be a life saver is an old station wagon. I found a 1963 Ford. The body was shot but the motor and transmission were in good shape. I have come to call it my "poor man's pick-up." If you price pick-up trucks you'll know why. An old station wagon can haul (one way or another) 90 per cent of the supplies a pick-up truck can. My theory was that the old station wagon couldn't be hurt any worse than it was so I loaded block sand, wood, steel concrete and anything and everything in it. If it has to go to the junk yard it was money well spent to have an "almost pick-up truck" for building this house without paying for and destroying a good truck. Give it thought.

Needless to say you'll need the basic tools. They all should be in good condition. In following chapters I will point out additional ways to save money as I discuss more specific stages of construction. Congratulations! You're doing a good job. By the way, how are the building inspectors treating you?

Codes and Regulations

Probably the largest single stumbling block you are going to face is how to comply with the local building codes. Once you face the facts that most building inspectors are long-term government employees and must constantly justify their existence you can easily understand why the mention of an underground house can start a panic. If they inspect and approve any phase of your building project and heaven forbid something drastic happened like a fire or a collapsing wall, they are a possible link in the responsibility chain. So therefore, they will take no chances of looking bad in their job performance. Consequently, they usually just overdo everything. They will stick strictly to the book. If it's mentioned in their code book, they aren't interested, because they are not liable. To prove my point, one regulation in our local codes covers hand railings. I have a set of four steps leading from the interior area down to the garage level. A hand rail is a must by codes for safety reasons. Any time you have three or more steps there must be a railing. Of course, I can't really say anything against the hand rail regulation except that it seems to be more of a nitpicking, harassing technique than of any real physical value. This code is now enforced diligently in our county.

In chapter 9 I describe how I built the support structure for pouring the concrete roof slab. You will have noted that in my roof alone, there are more than 140 tons of concrete, 10 tons of steel, 800 tons of dirt plus a two-ton dome. But to this day, no official or inspector has questioned the strength or method of construction, because it's not mentioned in their code books to do so. Pouring this roof alone could have killed as many as 15 people if it had collapsed in mid-construction, but officially no one cares. I say officially because it is just that.

PROFESSIONAL ENGINEERS

The official stance that my local inspection department took was that I must get a professional engineer to put his stamp or seal on my drawing. This seal I mention is a legal signature that indicates that the signing engineer has checked strength and material and agrees that the architectural drawing is satisfactory. This type of engineer is tested by the state for proof of his ability. Then he is given permission to charge a fee for checking these drawings, thus taking a share of the responsibility. So you see, as long as the engineer had his name on the drawing, the county inspection department could say that he was responsible if something wasn't designed correctly. To verify this fact, I can tell you that on the day I poured 3 yards of concrete in a footer, I had two separate visits from official sources, just to see if I was doing it right. Quite a paradox, but I'm sure you'll see it as you build. They are geared to conventional home building and not underground homes. I should also like to add that this is truer the more rural your location. If you were to build a strange structure in a district where high-rise office buildings, large industrial complexes and big apartment complexes are constantly under construction, the inspectors there have seen and are familiar with practically any type of construction problem.

On the other extreme, at the present time, there are still some areas of the United States that have no building codes at all. God bless their nonmeddling souls. It's a pleasure to realize that some communities still believe that the best government is the least governing.

If you intend to build in a community that has a complete set of building codes and zoning regulations, I'll remind you of a few things that will give you and them concern.

MEANS OF EGRESS

Probably the most difficult to comply with is a term called *means of egress*. This meaning exits. This covers doors, windows, holes or anything else that could be used as fire escapes. Usually the codes require two exits from each room except the bathroom. Obviously when you build underground the windows are the first to go, thus a problem.

Sprinkler System

The solution that was proposed to me was to use a commercial sprinkler system in place of the missing windows. If you price their systems, you'll soon find it unreasonable from a cost viewpoint. In my house an acceptable sprinkler system would have cost approximately $10,000. Needless to say, I couldn't accept that.

Smoke Detectors

Therefore, after much negotiations, conflict and letter writing, we (the county and I) agreed that a smoke detector in each room would be adequate. However, even on our compromises we ran into a snag. My idea of a good smoke detector was a battery-operated model made by any of the major smoke detector manufacturers. When I say battery-operated, I mean the type that doesn't require hook-up to the 120-volt house electricity. They only require small voltage batteries, usually one 9-volt battery that lasts for over a year. My logic for this type is understandable, I think. If the house electricity is out of order by storm or if a fire starts in the electrical circuit, it usually blows the fuse or breaker, thus shutting down the power to the detector that is supposed to be on guard. But a battery-operated model is on guard at all times. These battery models usually have a warning system that would sound when the battery is wearing down and ready for replacement. The county inspectors, as usual, had a different view of the same subject. They say that human nature, more often then not, will forget to replace the run-down battery so that the detector system will most likely be inoperative after the first battery wears down. I felt I was right, but I admit that it's all in how you analyze the problem.

Series of Corridors

There is another alternative to meeting egress (fire escape routes) regulations. That is by a series of corridors leading from each room (Fig. 5-1). The reason a corridor or hallway is required is because most local codes say you must be able to escape to the exterior by means of a window or door, without going through another room. For example, you cannot escape out of your bedroom into the living room to get to the nearest door or window. Your bedroom must have a window that leads to the outside and a doorway that leads to a corridor, since corridors are not considered rooms by definition. Now you can easily see why underground homes get into conflict when you eliminate all of the exterior windows. To meet all codes by the corridor system, you would have to waste a lot of square feet of expensive floor space, plus isolate some living areas. Corridors are not my personal idea of a solution, though you may work out an acceptable arrangement.

In summary of this fire escape problem, you have basically four choices to consider before you finalize your plans:

- Provide a corridor to each room
- Provide an appproved sprinkler system
- Provide approved smoke detectors in each room
- Provide escape from each room through the roof

It is my opinion that the third solution is a very economical and logical approach, since that is the way I solved the problem.

The fourth solution is listed more in gest than for serious consideration. I hope it's obvious that the more holes you have

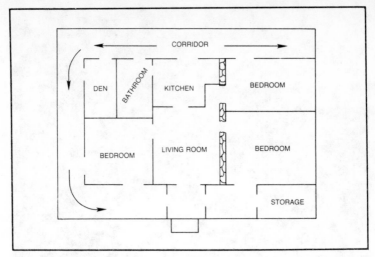

Fig. 5-1. Typical example of using corridors as means of egress.

in your roof, the more problems you will undoubtedly have. Also the cost could be prohibitive.

RETAINING WALLS

Another area that could cause you some trouble when trying to meet codes is retaining walls. Most underground homes have more than their share of retaining walls, depending on the design, of course. Most localities have regulations as to maximum height and method of construction. Try to arrange your design to eliminate exterior retaining walls, if at all possible. If you can't, figure the cost closely. They are extremely expensive to build if they meet the code.

If you expect a retaining wall to stand on its own when the force of mother nature, by means of freezing and thawing the soil, combined with the weight of the soil, is trying to push it over, then you are naive (Fig. 5-2). You can reinforce the wall with enough steel and enough concrete to insure that it will stand and not crack against any of the elements, but it is difficult. One of my solutions to retaining walls is to always brace one against another with steel-reinforced concrete (Fig. 5-3). If you have only one retaining wall, you can always tie it to an anchor post of some type (Fig. 5-4).

Fig. 5-2. A large retaining wall is unsupported in final stage of construction.

Additional suggestions on how to tie a retaining wall back could be lengthy and it is a subject that has been written about in detail. Ask a good contractor or engineer, or search the library for books on this subject. It will be worth your time to particularly look into the use of railroad ties. Most railroad ties which you can buy now have never been used on a railroad and they aren't really the same thing. The old-time railroad ties were pressure cooked in creosote (a preservative) to prevent rotting. The new so-called ties have only been dipped in a preservative and will not last as long as a real tie. Building retaining walls of ties is a difficult, back-breaking job. So use something that will stay in place as long as possible.

HAND RAILINGS

Another subject that will require close attention to complete your house is hand railings. There is usually a difference between safe conditions and meeting the codes. I ran into a problem with my interior garden (Figs. 5-5 and 5-6). When a garden is covered by a dome, it is a personal judgment as to whether it is considered exterior or interior. If indeed the

60

inspector feels it is exterior, then hand rail, doors and electrical boxes must meet one segment of the code. If he decides it is interior, then he will turn to another page in his book. As I mentioned, hand railing requirements are different depending on the use of your garden or atrium. If you plan to use an

Fig. 5-3. Two retaining walls held apart by the roof structure.

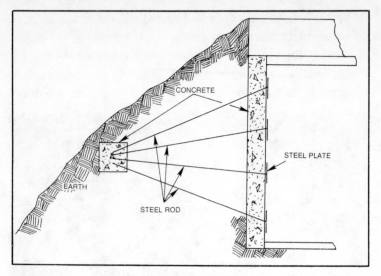

Fig. 5-4. Alternate method of holding a retaining wall.

above-floor level to grow year-round plants, then you are required to have a railing, even if you only use it once a year. If it is just for decoration, then no railing is required. Still another problem area could be glass doors. The codes sometimes say sliding glass doors are not allowed as a principal means of eggress. This becomes particularly complicated when these doors enter into your garden. Is it exterior space or interior space?

Most local building codes were patterned after a basic set of national regulations and sometime in the past these codes have been adopted as the gospel truth. In addition, many subdivisions have drawn up a few amendments to the national codes. Most of these amendments are pets of some local developer and in no way are intended to serve your benefit, only your expense. For this reason, everywhere in this country will probably take a different approach to the building of an underground house.

AIR CIRCULATION

Air circulation is one area that is touchy and important. If your house is like mine when complete, you will have only a wood stove for heat. I'm not suggesting that you don't put conventional heating in if you feel comfortable doing it, but the majority of underground home owners find wood heating is the most satisfying, even over solar heat. If you use conventional heat with a duct system, then all air circulation problems will

Fig. 5-5. Partial view of a garden in the final stage of construction.

Fig. 5-6. An exterior garden and walkway.

be easily met. But if a wood stove is in your future, then concentrate on the area of circulation. Just because you heat one room easily and quickly doesn't mean the hot air will move from room to room. You need a circulation system, but as you know a wood stove needs the same oxygen that you need to breathe. As the fire burns and you breathe, this oxygen must be replenished. So you do need a duct to the outside world. Here is where the codes will conflict with your good judgment about the size and location of the fresh air intake supply. The inspector's logic here is that in a conventional home fresh air is drawn in around windows, under doors, etc., regardless of the new caulking and installed insulation strips. I know from experience that all you need to do is circulate the air inside the house. The exterior doors would have to be opened approximately once a day to provide all the fresh air necessary to live comfortably. This is particularly true if you have a large dome or breathing skylight. Remember, if you keep bringing in outside air in the winter when it's not required, you defeat the purpose of underground living. Of course, the same is true in the summer. In summary of fresh air, I like and need fresh air as much as anyone, but enough is enough. Don't change air more frequently than necessary.

ELECTRICAL CODES AND PLUMBING

Plumbing and electrical codes will be mentioned only in passing. There doesn't seem to be much of a problem with building underground and meeting plumbing and electrical codes. Electric wires and water pipes don't know the difference so at least these two phases of utility installation should be simple (Fig. 5-7).

I will tell you to watch out for the sewer drain vent pipes. Don't put them through the roof. Run them with the water pipes along the block wall and vent them to the exterior vertical walls. These drain air vents are required to be 2″ in diameter to allow proper air intake into the septic system. The reason these drains are required is that if you release a large amount of water into a drain line at a certain point, it creates a vacuum as it drains to a drain field. This vacuum draws the standing water out of the traps in the sinks at another point. Once the water has reached the septic tank, the trap at the second point will be empty and allow sewer odor to seep back up the pipe into the living area.

STAIRS

Still another area that you might watch out for and design around is the steps. The reason to avoid steps in an under-

Fig. 5-7. Typical water pipe attached to 1 × 3 firring strip on block wall.

ground house is, once again, written in the sacred scrolls of building codes because a principal means of egress cannot contain steps upward, only downward. Therefore, since you are already down, you can't continue in that direction. You must eventually come up. This is a code that varies from locale to locale, so check your particular situation out carefully.

DOOR OPENINGS

This chapter will be closed with one last code for you to check on. Most likely all doors opening to all living areas from utility rooms, storage rooms, garages, laundry rooms or furnace rooms must be made of solid wood approximately 1½″ thick with no windows. They also must have an approved burn rate. The reasoning is that it will contain a fire in these work areas long enough for you to escape past (not through) the door.

As you go through these trials and tribulations, you'll find you cannot meet certain codes, or it is really impractical to do so. If that is the case, there is a means of appeal if the building inspector will not suggest an alternative to your problem.

BUILDING CODES APPEALS BOARD

This course is usually referred to the building codes appeals board. They are usually politically appointed members for a specific term. They have absolutely no power to change a regulation. For example, they cannot give you approval to eliminate a door if a door is called for by code. Most codes make use of the overworked and all-encompassing word *approved*. If you put a hand railing along a stairway and use a piece of wood 2″ × 2″ thick, the inspector may disapprove the railing. You can ask for an appeal to the appeals board saying that 2″ × 2″ is strong enough and should be accepted. Then they will rule either for you or against you. They will never say a 2″ is not okay but 2″ × 3″ is okay. This means that they will not suggest what is acceptable, only give you a yes or no answer.

6

Site Preparation

Site preparation for an underground house is definitely more critical than that of a conventional home since most underground homes are on the side of a hill. This terrain will be covered first and in most detail.

Chapter 2 covered what to look for and what to avoid when finding your ideal location. There is an interrelation between these two chapters. As you read on, you will notice that I am explaining location to you as if your land were in the two- to three-acre range. I think of this as the minimum size for an underground house. It is obvious that if you have 10 or 15 acres you should have little trouble locating the ideal spot for everything. Therefore, I will only point out the things to watch for if your land is in the two- to three-acre range (Figs. 6-1 through 6-3).

Once your boundaries are clearly and accurately marked by surveyors' stakes, find an observation point where you can see most of your property and study the contour.

DRIVEWAY

You also have to give a lot of thought to where you are going to cut your dirveway. This is the first excavation that

Fig. 6-1. A two- to three-acre lot.

usually takes place. I suggest you put your driveway on one side or the other as opposed to down the middle of your property (Figs. 6-4 and 6-5). If your driveway is in the center of your yard, you split the lawn into two parcels, both of which may not be big enough to use for recreation areas.

SETBACK

You must also know what the restrictions are on *setback*. Setback is a term that your locality has established to name a suitable distance from the public highway where you can build your house. This distance is different for each piece of land. The reason for this regulation is to prevent someone from setting a new house right up against the roadside. There is also a regulation controlling how closely you can build to your neighbors' property line. This dimension is usually 10 to 20 feet. If you have a three-acre tract to build on and you follow all of the possible regulations, you will soon notice that you don't have as much choice as you first thought concerning the location of your house.

Fig. 6-2. The minimum size lot for an underground home is at least two acres.

SEPTIC-SEWER SYSTEM

Before you start driving markers in the ground locating the corners of your house, have you cleared the location of your septic-sewer system with the health department? If not,

Fig. 6-3. Preparing your site.

Fig. 6-4. Cutting your driveway out of your land.

be sure to check with them. And don't forget the well is also covered by regulations. Usually the local health department handles this. Of course, I'm assuming that your land is covered by codes because most land near major population centers is heavily regulated, especially if it's a small parcel and not of farm size.

Fig. 6-5. Your driveway should be on one side of your property.

SURVEYING

Now, back to site preparation (Fig. 6-6). It's time to begin putting stakes in the ground. You can locate the corners of your house yourself with a transit, or you can hire a professional surveyor. If you do it yourself, follow this method. You

Fig. 6-6. The lot before excavation.

Fig. 6-7. The lot one day after excavation.

must have a set of drawings of your house by now. Add 10 feet to the width and length. The reason for this is that the stakes you drive in to the ground when laying a house do not indicate the actual exterior wall corners but the corners of excavation (Fig. 6-7). The bulldozer operator will use these stakes as a guide to dig the home. The purpose of the extra 10 feet on the length and width is that the block layer or form builders need room to walk and work around the actual wall once they are started (Fig. 6-8). Also the drain field is laid in this same space once the walls are started.

As soon as this same dozer cuts your driveway and you agree that it's the right location, bring in gravel for a driveway bed. Look in the yellow pages under *gravel* or *quarries*. Common terminology around Maryland is *crusher run*. This is a gravel and dust mixture that is as cheap as any gravel and packs to make a good road bed. You want this driveway passable for your delivery trucks right from the start because they will not deliver material if there is a chance of getting stuck on the job site. You can't blame them for that because their trucks are expensive and serve as their basic means of livelihood. Besides, you could be liable for towing bills or damage to the trucks.

Whatever you do, don't let anyone talk you into paving your driveway with blacktop or macadam the first year. Let the gravel pack and settle for at least one year. Of course, if you are going to concrete your driveway, that's a different story. Concrete could be poured as soon as the house is

Fig. 6-8. Four foot walking space and drain pipe.

complete as long as it's not over unpacked fill dirt. Don't concrete or macadam your driveway and let heavy delivery trucks drive over it the first month it has been poured. Give it a chance to cure. A couple of months should be sufficient. If it cracks after two months, it wasn't put down or reinforced properly in the first place. Then your problems are just beginning as far as driveways go.

Give some more thought to putting anything down on your driveway other than gravel. The tax assessors love people who improve their property by a hard driveway. It's not fair, but it's the system of tax assessment.

MOWING THE LAND

If your land is grown over with weeds or undergrowth, find a way to mow it to grass length while the excavation is going on. Every neighborhood has someone with a big field mower behind a farm tractor who will do this for a reasonable price. You'll want to actually see the surface of the land that will become your future lawn. If it is grown over with brush, I don't care how many times you walk over a piece of land, you never see everything. For example, big rocks that require a dozer to move may be just below the surface. Move them before the dozer leaves the site. It's impossible to move a rock as big as a refrigerator, even with the backhoe I hope you bought earlier.

The next reason for mowing the land is to eliminate rodents, insects and allergenic weeds. It will also allow the land to dry out.

If you have a wooded area on your property, try to cut down underbrush, dead trees and unhealthy trees to allow the stronger ones to develop into good shade trees. Talk to a professional regarding the selection of the trees you save as opposed to the ones that get the axe. Some may be rare or otherwise valuable. Some may be dying and unnoticeable. Some may be thorny or have poison berries or leaves. Don't go in and methodically cut down all the living trees and bushes, on the premise that you will have to buy from a nursery anyway when the house is complete. It's usually a waste of natural beauty and nursery plants are expensive and very difficult to start. Keep your underground house surrounded by a natural setting as much as possible. You'll be happier in the long run and so will your neighbors.

PROPERTY LINES

Another thing to do regarding site preparation, which I strongly suggest, is that once the official survey is made by a professional and the property markers officially set, remove any wooden posts and drive steel pipes in their place. Do this at least every 50 feet along your neighbor's property line on all sides. Drive them to grade level, but make sure you can find them if you need to.

Some wise man once said, "Fences make the best neighbors." This at first may sound like a cold or unfriendly attitude, but it is not. Personalities change, situations change and neighbors change, and there's nothing that will cause hard feelings faster than two people who think they both own the same rose bush on a common boundary. I don't necessarily agree with the fence theory, but at least make sure it's obvious where the property lines are, right from the start.

The reason I point this out is that with both homes I have built the question has come up as to whom is using whose property to plan proverbial rose bushes.

Due to the fact that you are building a "strange" home in the eyes of most people, the world you live in may not be ready to jump on your bandwagon or beat a path to your door. In fact, they could be downright nasty, once your intentions to live underground are made public. So cover all possible sources of contention, especially those boundaries. There's no use in asking for any extra unfavorable reaction.

As site preparation continues, have your dozer operator grade off approximately 6" of top soil to form a mound. This method will prevent wasting top soil that you will need for final grading. If you don't pile your top soil aside, it will be covered by the lesser-grade under soil dug from footers, foundation or the driveway. If you think the extra grading effort to save top soil isn't worth the extra money you might have to pay a dozer operator, you'd better check the price of a truckload of good top soil. You'll agree that it's easier and more economical to save even an ounce of good soil rather than pay to haul it in at final-grading time.

7

Preparing and Pouring the Footers

Preparing footers for an underground house is much more critical than those of a conventional home, because of the excessive weight bearing on them. The consequences, also, are more critical. Cracking walls will lead to leaks.

Dig your footers correctly and accurately and it will be time well spent. To begin with, a conventional house footer is usually required by code to be 16″ wide by 8″ deep. I recommend that for an underground house over 2000 square feet that you increase the width to 24″ wide and 12″ deep. Also add reinforcing steel rods as an added safety factor against settling cracks. The amount of steel rod added to the footer is immaterial. The more the better, but it is expensive. So use good judgment and ask for advice. In my opinion, four pieces of steel rebar 3/4″ in diameter would be idea (Fig. 7-1).

The important thing about preparing and digging footers is that they are level and square. Also, the corners are at right angles. The reason to keep the corners square is basic. All that weight pushing down on the footer is trying to push it into the ground further (Figs. 7-2 and 7-3). To exaggerate this point, suppose the footers were only 1″ wide and pointed on the bottom. They would act as a knife edge and would keep

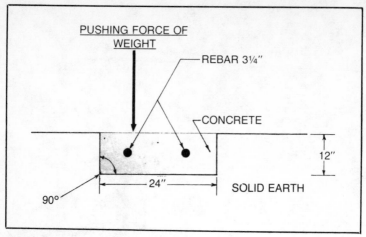

Fig. 7-1. Typical section of footer with rebar.

settling deeper and deeper. It's the wide, flat surface that stops settling in the same way that snow shoes stop you from settling in the snow.

This footer will be dug by a backhoe, either driven by a professional or by you. Either way the footers will be dug by a backhoe and cleaned up by hand labor and a square-tipped shovel. In preparation for the backhoe to begin digging, you have to place guide lines on the dirt for the operator to follow. Use a common field line that can be bought at any hardware store. Lay out the location of each footer edge (not center line) in the same way that football stripes are applied. The backhoe operator will then dig beside this white line you put down. Notice I said mark the edges of the footer. If your addition and subtraction is not accurate, you will be digging footers where there are no walls and vice versa.

Now that these white lines of lime are in place and the machine is digging, you need two additional helpers and a good transit.

Set the transit up totally outside the perimeter of the house foundation by at least 25 feet. This will prevent having to relocate the transit as the backhoe maneuvers closer.

As the backhoe is digging, you and a helper will continually measure the depth of his diggings to maintain a uniform ditch. Approximately every 4 feet drive a wooden or steel stake in the center of the footer. These stakes are driven in the ground until the tops of all the stakes are level with each other, as established by the transit reading. The stakes are spaced equally as the machine digs, and your helper takes readings on the transit. This continues until all the footers are complete.

Fig. 7-2. Typical 20″ wide footers.

Up to this point, the footer preparation is the same as in a conventional building. The big difference is the fact that every wall is a bearing wall. Therefore, every wall requires a footer underneath. This may not sound like a major point, but consider the fact that by putting footers under every interior wall, the amount of digging, hand shoveling, concrete, steel and so forth is approximately two and a half times as much as a conventional house. For example, my house, above the ground, would have required 260 feet of footer. By being underground it actually required over 570 feet of footer. So you see how the cost and labor can add up fast. This, coupled with the extra width and depth I suggested earlier, could cause your underground house footers to actually cost over four times that of a conventional house.

Do not take these footers lightly. If you do, you will not only waste time and money, but you could endanger the complete integrity of the structure.

THE CONCRETE

Pouring the fresh concrete is a job you can handle with a little help from your friends, whether it's footers, sidewalks or your roof. First of all, all concrete companies sell concrete by the cubic yard delivered. You are responsible for telling them how much to deliver, and you are responsible for the distribution of this concrete. To figure how much you will need for a specific pour, you simply figure the cubic yards of the area to be filled—length times width times height. These dimensions are usually discussed in feet and inches so be careful when converting to cubic yards. There are 27 cubic feet in a cubic yard. I have heard stories from the concrete truck drivers about do-it-yourselfers who forgot that they were buying a cubic yard instead of a square yard and found themselves either far too short of concrete, or with enough to concrete the entire neighborhood. Remember, when you are talking to the concrete company on the telephone, he has no idea what you're doing. So he takes your request to be the gospel truth, whether you ask for one yard or 100 yards. Be careful with the simple arithmetic. It could be costly if you make a mistake. Also remember that once the truck arrives with your order, it's your concrete. Concrete does not make round trips. If you have a footer to be filled that actually requires 5 yards, but through your mistake in ordering the truck arrives with 8

Fig. 7-3. Drive the stakes approximately every 4 feet into the center of the footer until the tops of the stakes are level.

Fig. 7-4. Accidents are caused by carelessness.

yards, you pay for your order and the excess gets dumped on your property for you to clean up later.

Once the truck arrives, the driver will know how close to get to your diggings. Let this decision be his. Most of them are experienced and are responsible for their trucks. However, I happened to get a driver who was a little more of a cowboy than a truck driver. Figures 7-4 through 7-6 illustrate the results of carelessness and poor judgment. This time the driver was only severely shaken. It's a miracle that this accident was not fatal, but the truck was totally destroyed. So let him make the decision as to how and where he will dump the concrete. When pouring concrete down a hole, the risks can be very great at times, especially when building an underground house.

Once the concrete truck starts pouring wet concrete in your footer, all you need is a good rake and shovel. The driver will ask you whether you want it wet or dry. He only means with more or less water added. If the concrete flows along the footer easily by raking, then keep that consistency. Put the stakes in the footer to insure the proper level for the concrete. Don't forget to watch them closely as the concrete flows. They are now the only guide you have to keep the concrete level enough to lay the block on. I'll remind you right here that the interior footers of an undergrund house need not be as

Fig. 7-5. Poor judgement caused this truck to be totally destroyed.

level as you may think, because a 4″ to 6″ slab of concrete gets poured over these interior footers. Don't try to pour footers and slab at the same time. It's a big job and there's no advantage to be gained except very little time. The best reason for not pouring footers and slab at one time is that if you have good drainage soil, it will be a sandy, mica soil and very loosely packed as opposed to a clay-type soil. It is often very difficult

Fig. 7-6. Fortunately, the driver of this mishap was only severely shaken.

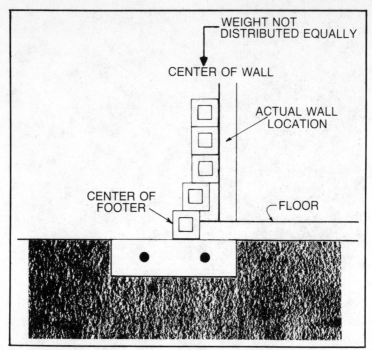

Fig. 7-7. An exaggerated method of misalignment.

to keep footer ditches from caving in along the edges. If the side of your footer ditch does cave in, you either lose strength or waste concrete. If you pour the footer and continue pouring more concrete to make the floor slab, the edges will most likely collapse. This is my opinion, and I'm sure some people reading this will disagree, but remember I'm living in my house and everything is working well.

Now back to the footers you're pouring. In most cases the concrete will flow freely 10 or 15 feet with little raking and shoveling. Once the raking becomes difficult, ask the driver to move to another spot if he thinks he can do so safely, and continue to dump until all the footers are filled to the top of the leveling pegs. There is no need to use a trowel on the footers. The surface, as it settles and hardens, is smooth enough as long as it is level to plus or minus a ½". If you misjudge and the footers set up out of level, the block layer can correct the mistake, but it is a slow procedure and most block layers hate to begin laying block on footers that are unlevel or out of square.

The principle of right (90°) angle corners is the foundation of basic building construction. The entire weight of your house plus the earth covering it rest on these footers. If by chance you poured them out of square, then the block layer will have to shift his first course of block to the edge of the footer to compensate for your error (Fig. 7-7). Once he does this, the weight is not distributed evenly and a crack in the footer, wall or even roof is likely to happen in the future. If you are not confident in your ability to keep the footers accurate and square, hire someone to help you. You can't end up with a successful underground house if the foundation is a weak link.

SEWER PIPE

Once the footers are poured and set up, check all the cast iron sewer pipe you placed (Figs. 7-8 and 7-9). This is the only time you will get a chance to do anything about mistakes or damaged pipes. The plumbing inspector has approved your system most likely, but that doesn't stop accidental breaking

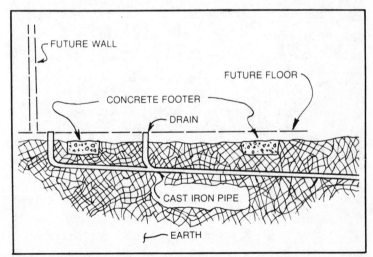

Fig. 7-8. Cast iron drain pipes are tunneled under footers. They are not placed prior to pouring.

Fig. 7-9. All pipes must be thoroughly checked.

or movement as final footer preparation takes place. The next chance you will get to correct a drain pipe will be at great expense, because you will be using a jack hammer to break the floor apart. Check the pipe; it will be a load off your mind later on.

8

Walls

There are only two basic methods for constructing the walls of an underground house, as opposed to the endless methods and materials available to conventional, above-ground home constructions. Above-ground construction lends itself to plastic, wood, glass, stone, brick, metal, concrete and any combination of the above, but in underground construction the material is limited to one type—concrete. There are two possible ways of erecting this concrete:

- Pouring concrete in forms
- Laying concrete block.

FORMED AND POURED WALLS

This method is the strongest, most difficult and most expensive. Needless to say, any advantage you gain you pay for. As I'm sure you are aware by now, there is always more than one way to do any job. The same goes for pouring concrete walls. Insert a section of steel into the poured footers immediately after they are all leveled out. These pieces of steel should stick out of the center of the concrete about 12″. So if you poured your footers 12″ deep, the steel rebar would have to be cut into approximately 24″ lengths. Put these steel pieces on 2 foot centers everywhere a wall will be. Remember to skip putting a post in the section of footer where a doorway will be. If you forget or misjudge, you can simply cut it off later. It's not that big of a deal. The reason for these steel posts is to tie the walls solidly to the footer to prevent any possible movement. This method also increases strength. This is the type of construction you would use if cost were not a major factor.

The alternative to tying the walls to the footers is simply to set the wall on top of the footer and rely on the actual bonding of concrete to concrete, rather than a steel connection.

Once you decide on the method of setting the poured concrete wall on the footer, the actual wall-forming preparation is the same. You can locate form builders or renters in the yellow pages of most major cities. Renting forms will be the easiest and the most economical if you can find them. Erecting these forms is a laborious job, and you should have someone experienced in form erecting to assist you. If you rent these forms, remember, you are responsible for any pieces that you damage or intentionally modify. Also there is always the chance of collapsing forms and the extra labor and expense used to clean up the mess. It just isn't worth the extra trouble and cost to pour exterior walls.

BLOCK WALLS

I'll state right here that it is my preference to use concrete block for all walls, exterior and interior, instead of poured concrete. Think about the pouring method, analyze it, then forget it.

Check around and get a respectable block layer and contract price for the complete job. This is where your good judgment comes in. No one can help you make a choice. Only you have the facts and you have to live with it.

A good block laying crew will take about 10 working days to put up an average underground house. It will then be ready to begin preparations for the roof. Figures 8-2 through 8-5 illustrate the different stages of block work. Construct all walls, even closets, out of block. Exterior walls should never be less than 12″ thick. They are just right at that width to put rebar in and fill with concrete for strength. A post of rebar and filled concrete should be created any place that an interior wall does not intersect with an exterior wall. For a distance exceeding 10 feet, these posts should be approximately 4 feet long or three block (Fig. 8-6). Use rebar about ¾″ in diameter and insert a length down each hole in the block all the way to the footer if possible. The interior wall can be 6″ concrete block and doesn't need to be filled with concrete except in rare conditions. The are several reasons why you don't fill interior block with concrete. First, cost is prohibitive. Second—your electrician will use the hollow block to feed some of his wire through to meet certain codes. The wires can only be fed to the recepticle box by way of the hollow concrete block. (Fig. 8-7). Third, and most importantly, you'll probably have to knock at least one hole in the interior wall due to oversight of some trivial dimension. Have you ever tried to knock a hole in solid concrete? Forget it unless you have a power air hammer. I'll tell you a few of the reasons I had to knock out block after

Fig. 8-1. Concrete forms in place for a conventional house.

Fig. 8-2. Early stages of block laying.

they were set in place. I forgot to leave openings to the back of the bathtubs to allow access to the plumbing (Fig. 8-8). Also, there had to be a hole for a vent pipe out of each bathroom. However, this I didn't forget. I planned to locate this as the construction progressed. Also, don't forget that the water pipes have to go through the walls to the kitchen and bath. In

Fig. 8-3. The mid-stage of block laying.

Fig. 8-4. Block laying continues.

Fig. 8-5. Final stages of block laying.

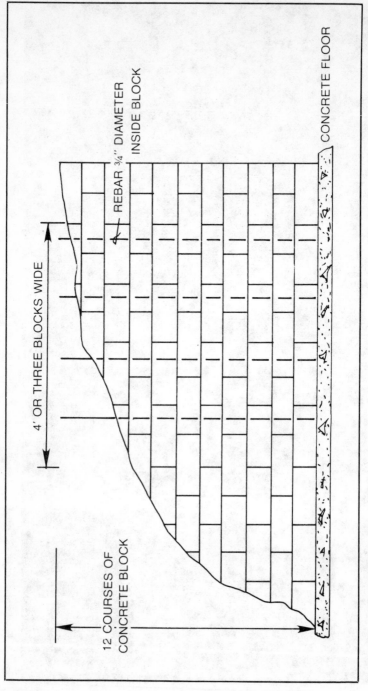

Fig. 8-6. Four lengths of rebar make a post.

94

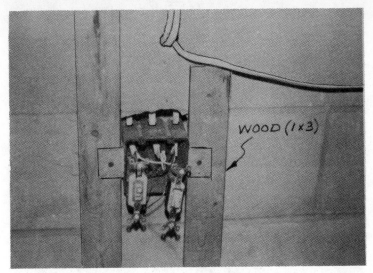

Fig. 8-7. Electric receptacle and wire in concrete block.

my house the copper pipes pass through the walls at approximately eight different places. This would be a major job if the walls were solid concrete.

Another place that I had to chip away the block was where I tried to install my one-piece fiberglass tub and shower unit. I found the unit to be ½″ wider than my door opening. This was

Fig. 8-8. Rough access hole to plumbing behind the bathtub.

an oversight that didn't cause a real problem because we just broke away the block as needed.

Instead of knocking block out for everything you should plan better than I did and install vents and heat ducts as the block is laid. This sounds easier than it is. Vent and air ducts sometimes have to be relocated because natural air currents are drastically affected by being underground. I found that where I thought air would circulate naturally, it didn't, and vice-versa. Thus, I had to relocate numerous openings in the concrete block. I suggest you try to locate the vent openings as the block goes up, but don't be surprised if you miss a few.

As your block work progresses day by day you will, for the first time, see a house developing before your eyes and realize you're really undertaking an enormous project. There's no turning back at this point, so you had better like what you see so far.

9

Roof Design and Preparation

Now that you have all the walls up and level at ceiling height, you come to another really critical part—preparing the shoring (Figs. 9-1 and 9-2), the substructure and braces for pouring concrete. What you are actually doing is building a platform capable of holding as much as 250 pounds per square foot. This includes steel, approximately 10″ of wet concrete and a worker smoothing this concrete. Holding wet concrete until it sets is like holding onto the proverbial burlap bag full of bobcats. Wet concrete wants to go everywhere except where you want it. If it finds a small hole, it will make the hole bigger and bigger until it breaks through. This method completes the roof in one pouring with no joints. Then there are no seams and no leaks. This is also the method that you can do yourself with a little luck and some help from your friends. Be realistic when evaluating your abilities to build your own shoring and substructure and pouring the slab. You may want to have a contractor do it. Doing it yourself is a back-breaking job. But you can save approximately $10,000!

Here's how I did my shoring (Fig. 9-3): laid rough cut lumber down on the concrete floor, wall to wall, every 18″. You can get this type of lumber at saw mills, usually very cheap. It's the strips that are cut from a log before they get to the center where the good boards are cut from.

I did not glue this rough lumber to the concrete floor (Figs. 9-4 through 9-6). It is not necessary to go to that extra time and cost. Check around for this wood. You'll find it. Later, when you're finished your house, cut this wood up and use it in your fireplace. I paid two cents a foot for random length over 8 feet long, and I picked this wood up to save money, rather than pay a delivery cost.

For your next step you will definitely need three or four people willing to work. Don't try to erect scaffolding by my method without sufficient labor, free or otherwise.

Once you have laid rough lumber on the floor and nailed a section of 2 × 4s along the ceiling height, stand a 2 × 4 upright under the first piece of rough lumber. Try to start in a corner.

With a few more 2 × 4s nailed between the top and bottom rough lumber, you can lay a sheet of ½" sheeting grade exterior plywood on the top. As your helpers hold this arrangement together, you climb up to the top of the block wall and nail through the plywood and the rough lumber into one of the upright 2 × 4s. You now have the beginning of your support. Your supports will be the same throughout the rest of the house.

If, at this point, you are considering leaving out the rough cut lumber, don't, because the results will be catastrophic. The rough lumber prevents the extreme weight of the future concrete from punching a hole through the concrete floor or through the ½" plywood. It is a must.

Now back to the actual erecting. Continue to set 2 × 4s upright, watching closely to keep them at 16" centers, maximum. If your centers are a little less than 16", that will do.

Continue this method until one room is complete. Be sure to have a good piece of rough lumber at each joint of the plywood, because this is the weakest point of the plywood.

The sad thing about this method is that you are constantly cutting plywood sheets. Therefore, they can't be sold later. But don't waste the plywood—save all the small pieces. They

Fig. 9-1. All interior walls are complete and the shoring is started.

Fig. 9-2. The process of shoring is underway.

will fit somewhere as things progress. One good thing in your favor now is that you can climb up and walk around on this plywood platform that you're building.

Before you buy these 2 × 4 × 8s, I suggest you try to keep the ceiling very close to 8 feet, because sheet rock and

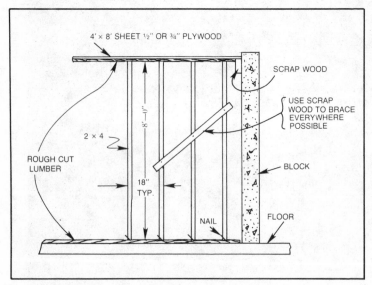

Fig. 9-3. A typical shoring arrangement.

Fig. 9-4. If you look closely, you'll see that the wood strips are not fastened to the floor.

Fig. 9-5. Wood strips are not fastened to the floor.

paneling come in 4 × 8 foot sheets. If your ceiling ends up 7' 11" high, you will have to cut each piece of sheet rock or paneling. That's quite a job. The way you control your ceiling height is by the length of the 2 × 4s. If you buy standard 2 × 4 × 8s and set them on rough lumber, add another rough board on top, both of which are approximately 1" thick and then set a sheet of ½" plywood on that, it's easy to figure that your

Fig. 9-6. Another view of rough wood and 2 x 4 supports.

ceiling will then be 8′ 2½″ tall when these are removed. That's too tall. Cut 2″ off each 2 × 4 × 8 before you use it. If you do, you will end up with a ceiling approximately 8 feet in height. Whatever you do, don't try to use precut 2 × 4s. Precut 2 × 4s are approximately 2′ 8¼″ long. This would make the ceiling too low.

Each room is shored up and braced individually. Do not let the wood go over the top of the black wall (Fig. 9-7). This method locks all the walls in place with concrete. In addition, it

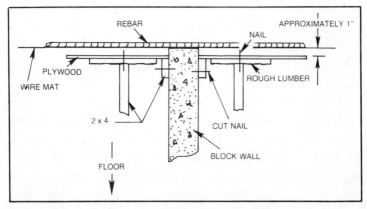

Fig. 9-7. The wood should not go over the top of the block wall.

holds all of your shoring secure from moving left, right, forward or backward. Most importantly, the rebar lays on the top of the block, thus avoiding an extreme load on the shoring. Once the shoring is up in all rooms, go back and check each 2 × 4 to make sure it is plumb and toenailed with at least two 10-penny nails, top and bottom. Remember one slip of a 2 × 4 and down comes the wet concrete. I can't warn you enough about this. A slip could bankrupt you. If you add the cost of the concrete, the cost of the lumber, your labor and the cost of cleaning up wet concrete once it has fallen, you can easily see why this could be the only mistake you'll make because it will also be the last.

Now that all the supports are moving toward completion and you can walk around on this platform, you're ready for the next step.

Stuff each open concrete block with paper. Save the concrete mortar bags from the block layer. Tear them in half in each hole of the concrete block to prevent concrete from flowing down the block. Filled block are too expensive and unnecessary for strength. Make sure your engineer agrees that all walls do not need to be filled with concrete. Some probably will, but don't fill any walls that aren't necessary.

Now your shoring is up completely, all the holes are stuffed with paper and you are ready to spread plastic. This plastic is to prevent the concrete from sticking to the plywood so that it will come down easily and will be resaleable. Use construction grade plastic that comes in big rolls from any building supply store. Cover all the plywood with one sheet of plastic. If you tear a small hole in the plastic, don't worry because a little concrete on the wood won't hurt anything.

There is no doubt that the most critical part of your underground house will be the roof. Put your time and money into the design of the slab. It will be money well spent. A good professional engineer should look your plans over, make suggestions and lay out a steel reinforcement bar pattern for you for a small fee. About $200 would be fair. When working up the original floor plan, be sure to keep the rooms near

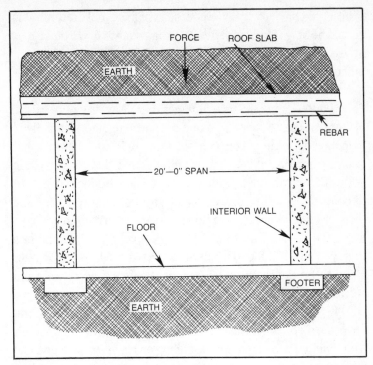

Fig. 9-8. Maximum span should be 20 feet.

conventional size. Since your interior walls are bearing walls and physically holding up the roof and dirt, you want to avoid spanning a long distance. The method of construction, cost and the reinforcement involved then becomes more critical and a bigger job than the average person can handle or afford. The widest span I recommend is 20 feet (Fig. 9-8).

A rule of common sense I would suggest for you to follow is that when laying out a floor plan, add the length and the width. The total should not exceed 38 feet. For example, a room which is 16 feet by 20 feet would equal 36 feet, or a utility room could be 25 feet by 8 feet to equal 33 feet. Remember this is only my rule of thumb. Trust your engineer as a final authority.

If your living requirements are average, 20 feet is as wide as any room usually needs to be. The engineer designing the strength of your roof will need an exact layout of your floor plan. If you have ever had a course in high school drafting, you should be able to make all the drawings necessary to build your house. If you don't feel comfortable doing this drawing, your engineer will have someone make a reliable set for you.

To locate an engineer, look in the yellow pages of the phone book under *construction engineer*. Any city of reasonable size will have a listing of one or more engineers capable of helping you. When you contact him, tell him exactly what you're planning and ask him what he would charge to provide you with a drawing showing rebar size, location and concrete thickness. Talk with this engineer at great length concerning the facts and details of your house. All the information you give him must be very accurate. For example, you can't tell him that you're planning on 5 feet of dirt over the roof when, indeed, this depth might be 10 feet. You must be specific and accurate with information.

Figure 9-9 is a sample drawing. Once again, check around for rates if you have the choice. This price will vary quite a bit from engineer to engineer depending on his interest in unusual dwellings, ecology or your plans in general. These are professionals and if they can possibly give you a break on price, they will, especially if they know you plan on building this structure yourself. They make up any financial break they give you on the next multi-million dollar project they design.

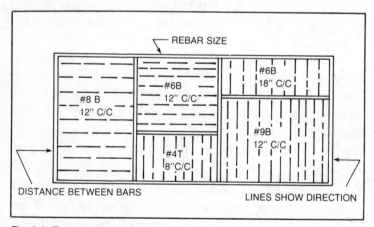

Fig. 9-9. Typical rebar drawing.

Another place to check for a good engineer is your local college. Some of those professors are sure to be engineers or have an engineering background. Before making a commitment to any engineer, check with the building code department to see if a professional engineer's stamp is required. I am not familiar with any locale in which a stamp or seal is mandatory for a private dwelling, but check to be sure. It will definitely cost more for an engineer to put his official seal on your drawings because this means he is legally responsible for the design. Just make sure the person figuring the strength of materials in your roof is qualified and interested in underground homes.

REBAR

Rebar is steel rod approximately 40 feet long and comes in diameters from ⅜" to over 1". These bars or rods are for strengthening concrete and nothing more. Your roof will probably use all diameters and lengths, depending on room widths. This is where your cutting torch comes in handy.

This cutting torch is by far the easiest and most universal method for placing, bending and cutting rebar because by heating the steel, you can make the bends that will be required in some places. There are other methods of cutting this rebar. You could use a circular saw with a metal cutting blade. This method is slower but still gets the job done. However, the bending will now have to be done by brute strength—yours. When large companies lay rebar, they have long-handled shears or cutters like a pair of bit-cutting pliers that will actually shear through some thicknesses. You, as a small contractor, may have a hard time locating this tool at a reasonable price.

This rebar will be delivered to your site in long lengths, which you will have to cut and bend to form the pattern designed by the engineer. Rebar can be bought from a used steel dealer listed in the yellow pages. The term "used" is misleading, because it's not used, it was simply previously owned and left over from a big construction project. Buy it if possible. Used rebar is about half the price of new rebar.

WIRE MESH

Another type of steel used for added strength in concrete is called *concrete mat* or *wire mesh*. It's a roll of heavy steel screen resembling a roll of fencing. It is manufactured in many sizes. This wire mesh adds additional strength to prevent cracking. I covered my entire roof with steel mat (Figs. 9-10 through 9-13), then the rebar was put in place. Discuss wire mesh and rebar with a couple of experts, especially the technique of tying rebar and steel together. When rebar is laid in a criss-cross pattern, it must be tied together with a piece of wire to prevent the steel from rolling and shifting as you walk on it (Figs. 9-14 and 9-15). Common stove pipe wire is often used. You must consider that wet concrete will be dropped on this steel from approximately 4 feet above. That's a great deal of weight trying to move your rebar around. It is critical that rebar does not move. Take your time and do it right. As the rebar is being laid in place, it cannot be touching the wood scaffolding. It must be raised off of the wood by at least an inch. See your engineered roof design for the exact height. This is so that the concrete flows easily under and completely surrounds the steel. There are wire pegs sold for this purpose, or you can use pieces of concrete block. Either way works fine.

Again, placing this rebar is most critical. It is physically impossible for one person to drag a 40 foot length, ⅝″ in diameter piece of rebar around on top of the scaffolding. You will need an extra person to control the steel. Once you have maneuvered just one piece of this rebar around, you will know what I mean by it being difficult to handle. Only experience will make it easier. Last, but not least, be careful as you cut through a length of steel. If it happens to be laying on an uneven surface, such as the ground, and you cut through the steel, one piece or the other could spring up and hit you in the face. The reason I tell you this is that that is exactly what

Fig. 9-10. The entire roof is covered with wire mesh.

happened on my job. The steel looked very relaxed, but when the cut was made, one end sprang up and caught me in the face. It could have been serious, but I was lucky. You might not be.

STEEL PLACEMENT

Now you have worked your way to the steel placement phase. Anyone can do this just by using common sense and following the instructions of your engineer or concrete expert. I fenced in the perimeter of my roof with block to form a totally closed area for pouring concrete. Once you get this far, you're at the critical point of your project. Don't let your wood frame work become exposed to rain, wind, etc., for a long period of time. Constant wetting of the plywood will cause it to separate and buckle, thus losing some of its strength. The sooner you pour concrete on the erected shoring, the better.

At this point you'll have all shoring in place, all holes filled and plastic in place as well as steel in place and raised off the plywood by at least 1″. It will also be laying on the blockwork. Finally, you must have a solid level that is an easily accessible spot immediately adjacent to the building for the concrete trucks.

As an alternative for shoring, instead of buying a truck load of plywood and 2 × 4s and building your own shoring, you could rent all scaffolding from a forming company listed in the yellow pages under *concrete forming*. They do not use the 2 × 4 method. They are very professional and use wood beams like railroad ties and a series of jacks. The catch to renting is that it's expensive. To rent all forming for the roof of my house for one week would have cost over $2,000. Plus I would have to pay for all of the plywood to be cut, and that adds up fast. Since most of the room sizes are not in intervals of 4′ × 8′, this requires you to cut numerous pieces of plywood. For this reason I shopped around for a volume price on plywood and 2 × 4s, bought them new, used them and sold them used through the newspaper at two-thirds the original cost.

Fig. 9-11. Wire mesh adds additional strength.

Fig. 9-12. Wire mesh prevents cracking.

Either method will work. You're the judge of your own ability and your accessibility to free physical labor. Okay, now back to pouring your roof. Just as everything else, there is more than one way to get concrete from truck to roof. The first way is a concrete pump. You can rent this equipment for about $300 a day. A concrete pump is just that. It pumps liquid concrete over distances up to 60 feet away. The problem with a pump is that it's slow and you need someone to hold a hose constantly at approximately every 10 feet of length because the throbbing effect of the pump tends to make the line uncontrollable. The second method, which is the recommended one, is a crane. They rent for about $300 a day, with one operator. The choice is yours. Discuss it with the local experts, especially the concrete finishers in your area. If you use a crane, it will take approximately six hours from start to finish for an average house. The crane will have two buckets. The concrete truck backs up to one bucket and fills it. Then the crane lifts that bucket to the farthest corner and dumps it. While this happening, the truck is loading the second bucket. By the time it is full, the crane will be returning with the first bucket, which is now empty. Keep repeating this sequence

Fig. 9-13. Wire mesh is another type of steel used for added strength.

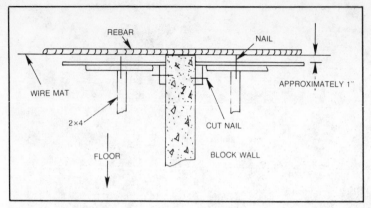

Fig. 9-14. Tying rebar to the mesh.

until the roof is covered to the depth you have pre-established (Figs. 9-16 and 9-17). As the concrete is dumped, you need labor (and lots of it) to level it and a finisher to smooth it. The smoother the surface, the less moisture the concrete will absorb when covered with dirt. However, don't use the buffing machine such as you will use on your floor. There is no advantage in getting the roof as smooth as your floor. One other thing to remember is to have someone shake the steel lightly by any method possible. You could use a pick as the concrete flows across your steel network. This insures that the concrete reaches all cracks and crevices and completely surrounds the steel. Concrete must totally surround all steel. The minimum thickness under the steel should be 1".

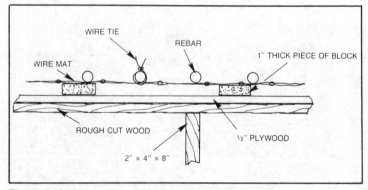

Fig. 9-15. Typical section showing rebar tied to the mat.

Fig. 9-16. A bucket of concrete is moved about by a large crane.

DO NOT use an electric or hydraulic vibrator to settle the concrete. It's not necessary and causes tremendous stress on the shoring. If your experts suggest a vibrator, watch out.

One hour after the last bucket is dumped and raked, you can relax. If it hasn't collasped by then, it will be there until

Fig. 9-17. A bucket of concrete is poured on your roof.

dooms day. You are now halfway to completion of your geothermic house!

ALTERNATIVE METHOD

There are other ways to accomplish the same goal. Here's an alternative.

If once again you check the phone book under concrete products, you'll find precast or prestressed concrete. There are companies that make slabs of concrete predesigned to carry any weight you require. These slabs usually have a maximum length and width somewhere around 20 feet long by 4 feet wide. As you can figure, this gives you quite a few joints that are sealed in a variety of ways, but are almost always covered with a second layer of poured concrete or sprayed insulation, such as foam. These companies will set these precast pieces in place of your walls for a price. The big advantage of this system is that there is no scaffolding to erect. It is rigid and it can be placed in one day. Of course, the disadvantage to this method is cost and the possibility of leaks.

Discuss all aspects of your roof with professionals. Talk to more than one person in each trade. They can all give you valuable information, but in the end you have to make the final decisions as to how and who. See Table 9-1 for some estimated costs of a roof.

Table 9-1. Cost of Roof.

METHOD	*ESTIMATED COST	ADVANTAGE	DISADVANTAGE
BUCKET AND CRANE	$8,000	FAST	NONE
PUMP	$8,000	IF CRANE UNAVAILABLE	EXTRA LABOR; SLOW
PRE-CAST	$14,000	ONE-DAY INSTALLATION; LESS LABOR	EXPENSIVE; SEAMS TO PATCH, POSSIBLE LEAKS
DIRECT DUMP	$7,800	NO RENT OF CRANE OR PUMP	LOTS OF LABOR AND WHEELBARROW; SLOW
CONTRACTOR EVERYTHING	$20,000	NO RESPONSIBILITY TO YOU	COST

*ESTIMATED COST FOR 3,600 SQUARE FEET INCLUDING ALL FORMING, PREPARATIONS, ETC.

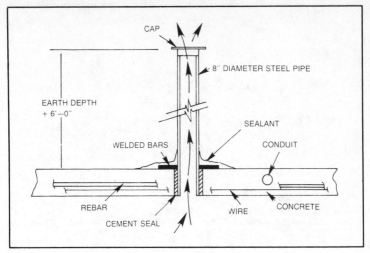

Fig. 9-18. Application of electrical conduit.

CONDUIT

It is a very good idea to lay down P.V.C. conduit (approximately 2″ diameter) on the steel before the concrete is poured (Fig. 9-18). This is for the main runs of your electrical wires. Be sure to let the electrician that is working with you locate and install this conduit. The codes are once again very tricky. The reason this is an alternative is that if your house is small, say 40 feet by 25 feet, the cost of putting pipe in the ceiling concrete would probably be more expensive than running wire around the walls. But if you want hanging ceiling lights, you should definitely give the conduit consideration. I put conduit in the concrete because my house is 90 feet long and 40 feet wide and I wanted ceiling lights in some rooms. This installation is not hard to do.

CHIMNEYS

After all rebars and shoring are in place, locate exactly where you want the chimney hole to be. Secure to the plywood anything you find that is 8″ to 9″ in diameter, maybe an old bucket or can. This will leave a hole in the roof for your future chimney. It is ridiculous to jack-hammer a hole in the roof after the concrete is poured and set up.

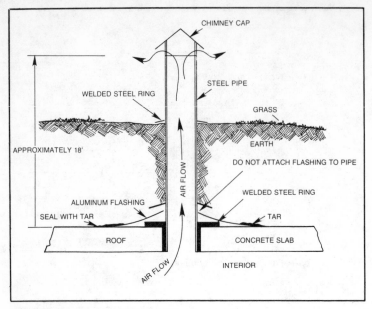

Fig. 9-19. Typical chimney installation.

As for building a chimney, in my opinion there is only one method that is solid and practical. Buy a length of approximately 8″ diameter steel pipe however high you want your chimney to be and weld bars to the sides (Fig. 9-19). These bars rest on the concrete and hold the pipe upright until you fill around the pipe with concrete and tar it over this new seam. It's that simple and works well. Standard chimneys will crack immediately after the first freeze because the block above ground level freezes solid and is forced to shift. However, the block 4 feet underground does not freeze and is mortared stationary to the roof which doesn't move, thus causing cracking. Watch this chimney phase closely. You can't live underground without one or even two.

Do not put any holes in the roof except for the chimney. Holes are just too hard to keep waterproof. Any venting can be put through the exterior walls just as easily as through the roof.

10

Waterproofing

When you mention underground homes or subterranian structures of any kind, people immediately and always think of a damp and dark hole. I'm sure their experience with older buildings used for storage or warehousing, and unfinished house basements with inefficient drainage and lighting are the culprit for this bad reputation. This is rightly so since many times the drainage is inadequate due to poor grading or planning.

Just as almost everything in the world has changed in the past 25 years, so have building materials, methods and equipment. For one example, urethane foam used for insulation was unavailable to the general public as recently as five years ago, and polyethylene and styrofoam have become commonplace for home construction.

Before going any further, I'll give my definition of waterproofing. It's simply preventing excessive damaging moisture from reaching the interior of your home. This can be done in a variety of ways with a variety of building materials. First, I'll give some examples of extreme methods to insure a moisture-free house. However, remember that no matter what you do or how well you do it, there's always the possibility of a moisture problem requiring mechanical assistance, such as pumps or at least dehumidifiers. I just want you to be aware of the potential problems. It's a risk you will have to take when building underground.

MOISTURE TREATMENTS

Now for the methods. Once you have excavated the land to your desired level, poured your footers and are ready for the walls, you may realize that you could have poured solid cement reinforced walls instead of concrete block walls. Solid concrete is a superior barrier against running water only. Whatever your exterior walls are, they should now be treated to prevent moisture vapor from penetrating to the interior.

Tar Baby

I used two heavy coats of hot tar, sprayed on by a commercial *tar baby*. Tar babies can be found in the yellow pages of the phone book under waterproofing—tar. Two coats of tar are sprayed on. The second coat is sprayed on after the first has dried. This method is far superior to one heavy coat. Another reason in favor of the second method is that any high spot or edge will only absorb a thin layer of tar. Additional tar will only settle to the low spots leaving the high spots only lightly covered. By spraying one average coat everywhere, letting it dry and spraying a second equivalent coat a day or so later, the build-up is uniform. This is the most common method.

Polyethylene Layers

If you want to go to additional expense, you can wrap the entire structure with two or three layers of polyethylene (Fig. 10-1). You can buy polyethylene in 100-foot rolls by 20 feet wide and four milligrams thick (about the thickness of a sheet of paper) from most building supply stores. Once the tar is on the block and dry to the touch you can begin to wrap the building, ending with two or three layers over the entire structure. Try to keep the polyethylene smooth and wrinkle-

Fig. 10-1. Polyethylene used as wrapping.

free. I used polyethylene sheeting but found that the wrinkles were a real problem because a mild breeze always seemed to be blowing. This proved to be more trouble than it was worth. Just for information's sake, polyethylene sheeting will not deteriorate underground but sunlight will deteriorate this material. Tests indicate that polyethylene that has been underground is still intact after 20 years. The only reason I am reluctant to urge you to use polyethylene is if you tear one hole in it as you back fill the dirt, you will lose most of the potential waterproofing value.

Pressed Insulation Board

Following the wrapping of the structure, you can begin to backfill gently (Figs. 10-2 through 10-6). Take care not to tear the plastic if you do use it. If you want to take another precaution, buy *pressed insulation board*. They are usually about ½" thick. It is as cheap as any building material you can buy in 4 × 8 sheets. Line the exterior walls with this material The only purpose it will serve is to cushion the rock and dirt as it falls against the polyethylene wrapping. If your soil and drainage is good this should be all you need. If you want to go to real extremes and you have an indication that your soil isn't as good as it should be, this next method should work for you. But it is expensive.

The most common method to keep water away from concrete floors is a series of pipes covered with gravel, leading to a drain line away from the house (Fig. 10-7). The less drastic method is to lay a couple of layers of the polyethylene down before pouring the concrete. These polyethylene sheets were the only prevention I took to stop moisture from penetrating through the floor and it seems to be doing the job. In reality I don't think anything would usually be required if the soil is of a good drainage quality.

Clay as a Water Barrier

One more thing you can do to help insure the proper runoff of excessive water is to have clay soil trucked to your

Fig. 10-2. First stages of backfilling.

Fig. 10-3. After the plastic is laid, you can begin the process of backfilling.

Fig. 10-4. Looking over the backfilling job.

Fig. 10-5. Backfilling continues.

Fig. 10-6. Both backfilling and excavation require heavy machinery.

Fig. 10-7. The land is dug out for a drain line.

site from wherever you can find it and grade from 6″ to 1 foot of clay over the roof (Fig. 10-8). Good quality clay will have a consistency similar to modeling clay. It should be graded to a peak and another 3 or 4 feet of good top soil added on top of the clay. Then if water happens to seep that far down, the clay barrier will cause the water to divert off the roof.

The condition you can create with clay by forming a pitch to the top of the concrete roof slab to divert water away from any opening can be formed into the concrete slab as it is poured. It is possible to pour the concrete with a pitch suitable enough to insure that water cannot flow toward an opening such as your garden or chimney opening. At the same time you could lay standard perforated drainage pipes in a bed of gravel directly on top of the concrete.

Use whichever system you feel safe with and top it all off with a crop of excellent quality grass. Once a good sod base has developed, keep it manicured smoothly and cut as short as reasonable. Follow your local landscaping expert's advice as to the length you can cut the grass depending on the temperature, rainfall and other conditions affecting the ability of your grass to survive. I mixed two grass seeds together in a 50:50 ration. One seed sprouted in three days when watered, but was an annual. This was only to prevent erosion until the permanent seed caught hold.

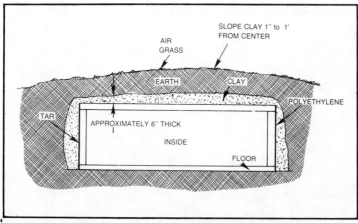

Fig. 10-8. Clay could be used as a natural water barrier.

SMOOTHNESS OF ROOF

One additional fact about waterproofing deals with the smoothness of the roof's concrete. The rougher the surface, the more moisture the concrete will absorb. If the slab is polished smooth, water will not penetrate the surface. Obviously, there is a happy medium to work for.

11

What About Moisture?

Whenever the subject of an underground home comes up in conversation, the same questions consistently pop up. What about moisture? Doesn't the house smell musty? Aren't there drops of water on the ceiling? Why isn't an underground home just like a basement? These questions are definitely logical, so I'll try to explain why an underground house has no real moisture problems.

Everyone has been in an old house with a clammy-feeling basement at one time or another. They have a problem because the waterproofing applied to many conventional home basements is only one step better than nothing at all. Then the grading is quite often done with mainly cosmetic results in mind rather than keeping water away from the foundation. Also, the top of a basement is exposed to a fluctuating temperature. In addition, the absence of activity in a basement causes a lack of the required air circulation. These are all reasons why a basement smells musty. The reason my house doesn't have these problems is basic. My grading was given top consideration for maximum water runoff. Then my waterproofing was applied cautiously.

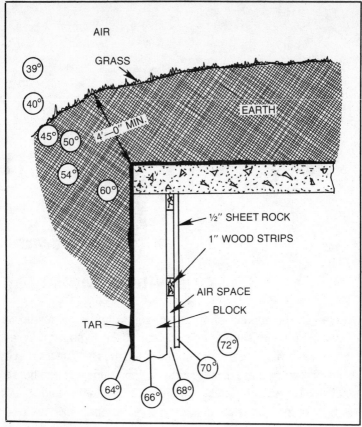

Fig. 11-1. Very slight temperature differential.

CONDENSATION

Once these physical moisture barriers are complete, the remaining moisture problem is the result of the law of physics. Condensation is the direct result of temperature differential and the percentage of humidity of the outside air. The temperature differential I'm talking about is best exemplified by what happens when the windshield of your car fogs up when you first start to drive when it's cold outside. The reason fog appears on the interior of the glass is that your body heat and breath are warm. Approximately ¼″ on the other side of the glass is a temperature of probably 50° or colder. The ¼″ thick glass is the point to consider. Remember hot and cold separated by a thin membrane will cause moisture to form on the warmer side.

The reason my walls do not sweat can easily be illustrated. The temperatures shown on Figs. 11-1 through 11-3 are estimated to help explain the point.

In Fig. 11-1 you will note that the temperature differential isn't present. However, Fig. 11-2 shows how great the temperature differential could be in a conventional basement. As you see, it is a similar condition to the cold car example.

In short, if the temperature is similar on each side of the walls, the remaining moisture problem can be handled with a standard room dehumidifier. The reason I don't have a problem around my doors is that the living area is separated from the outside temperature by a foyer arrangement (Fig. 11-3).

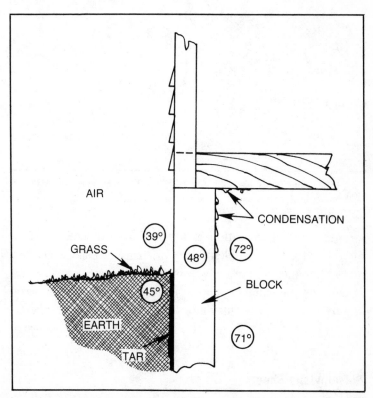

Fig. 11-2. Conventional house showing extreme temperature differential in degrees Fahrenheit from inside to outside.

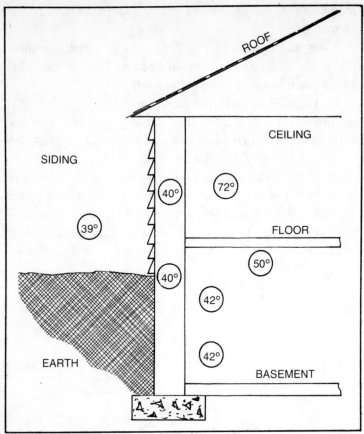

Fig. 11-3. Living area is separated from outside temperature by a foyer arrangement.

SPONGE EFFECT

Just as condensation is a potential problem, there is another moisture problem. It is technically called *capillary draw*, but more commonly it is called a *sponge effect*. The reason for the name is simple. Just as a dry sponge absorbs water, so does the air inside of your house. However, the air in your house is absorbing water from the earth because warm air holds more moisture than does cold air.

Polyethylene Sheets

There is, however, a couple of methods to prevent the sponging of humid air, or at least to help slow it down. The first

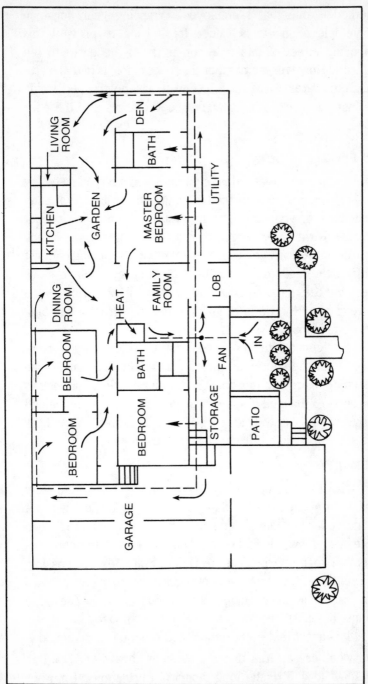

Fig. 11-4. Dotted line shows air movement by duct system.

135

method is one that I employed. Wrapping your house in polyethylene sheets is valued for helping to prevent this sponging effect of moisture coming from the earth to the interior. However, remember that once a hole is in this material, the water will definitely find it and be trapped inside against the block rather than be absorbed back into the drier earth.

Air Pockets

The second method is one that if you have followed my building suggestions up to this point, you will have already included it in your house. This system is the 1″ air pocket created by the wood furring strips used to hold the sheet rock up. Air pockets are the best way to prevent sponging of moist air from the earth to the interior. A 2″ air space would be more ideal, but the cost of furring all the walls with a 2″ strip of wood instead of a 1″ strip is probably prohibitive from a cost and labor standpoint.

The construction of my exterior and interior walls, combined with the special air circulation system I installed, are adequate to prevent most major moisture problems, as long as the air is exchanged frequently in corners, closets and behind furniture. This air will move about with the simplest of ventilating systems using a mechanical in-line fan.

HUMIDITY

However, remember that I am warning you that humidity will be a slight problem that you will have to deal with. A standard large room size dehumidifier will remove the extra humidity. Now, don't be shcoked by that statement. Remember that anywhere in the United States except possibly the southwest, people use dehumidifiers in their conventional houses. I'm just reminding you that you can expect the same simple humidity problems as in any other house.

It is a fact that each section of the country has conditions and weather patterns that are uniquely their own. For this reason and due to the long technical calculations, I am not going to go into great detail about how to figure heat loads, humidifier sizes, etc. However, there is one term that should be mentioned. That term is *dew-point*. This is the temperature at which moisture droplets will form. This is not a constant point. It fluctuates with the percent of humidity in the outside atmosphere. You learn a great deal about dew-point as you discuss the construction of an underground house. After lengthy discussions with numerous experts in the field of air handling, I proceeded to install an air handling system. Dew-point is another law of physics that applies to warm air holding more moisture in vapor form than colder air. This explains the lessening of a humidity problem in the winter.

Another thing you can do to lessen the possibility of a moisture problem is to insulate hot and cold water pipes. Also, insulate the hot water heater.

Moisture is a problem, but it can be conquered, probably as easy as any other problem connected with an underground house.

12

Backfilling and Grading

By the time you get this far you will have approximately 75 per cent of your money invested, listened to plenty of ridicule and put in hundreds of hours of hard work. Don't make a major mistake now. Whatever you do, don't do any grading near the building for at least three weeks after the concrete roof slab is poured. Even then I suggest you tread softly until the concrete is cured for five weeks. At this point I'll tell you of a change I will make on the next underground home I build. I will not try to waterproof the entire structure at one time. Note in Fig. 12-1 that I did not backfill against the exterior walls until the roof slab was complete. My reasoning for this was to keep the complete surface, four walls and roof exposed so that they could be sprayed with a hot tar-like sealer and then wrapped in the sheets of polyethylene as previously discussed. This cavity was approximately 3 feet wide and 15 feet deep. It was constantly breaking away and then needed to be hand shoveled out. Try throwing a ton of wet dirt shovel by shovel 15 feet in the air over your shoulder and you will soon see that wrapping the structure in one piece isn't worth the effort. If I had it to do over again, I would complete the exterior walls and spray them with two coats of tar to within 1 foot of the roof slab line.

After the wrapping of the plastic and about one day of drying, I would begin pushing dirt against the wall, first checking to see that the drain pipe at the base isn't disturbed by the first load of dirt falling from as high as 15 feet. By backfilling as I suggest, not as I did, you will eliminate four of the major problems I had all during early construction. The first problem, as I said before, was the constant collapsing of the dirt before I had the tar sprayed. Secondly, it was a constant safety hazard and more than once I dropped tools into the hole and had to systematically climb down to retrieve them.

This is another good place for me to explain the inconsistency that abides in the bureaucracy of any department of permits and regulations. Follow this logic, if you will. As I was grading, an official stopped by to tell me that I needed a barrier of some type to stop soil from washing into a small stream nearby 300 feet away. Now this was and is a hillside of natural vegetation of underbrush, trees, bushes, etc., but he wanted me to provide a barrier of bales of hay or straw staked to the ground to prevent erosion into the small stream. The part that irritated me was that this excavation was never mentioned as a safety hazard to human beings. It bothered me, so I eventually put up a make shift warning fence. But the inspection officials never mentioned it. Another paradox of bureaucracy.

By doing the grading at two different times the dirt against the wall has time to settle, especially after a heavy rain. Do not do any of the grading until all interior block walls are tied in place to the exterior walls or these exterior walls will crack or collapse. If you now fill the dirt against the exterior wall, it will be easier to move around while preparing scaffolding for the roof pouring.

TRACTOR-SPREAD

Now begin to spread the dirt over the roof for curing unless your engineer says otherwise. Leave the underside shoring up until you have at least 3 feet of dirt spread smoothly over the complete structure. Do this with a small tractor, initially, similar to the one in Fig. 12-2. One like this can usually

Fig. 12-1. View of roof prior to start of grading.

be rented by the day or week. The reason I say a small tractor and 3 feet of dirt is that the dirt transfers the weight of the tractor uniformly downward onto the roof slab as opposed to the weight of the tractor being distributed to only the four wheels of the tractor. Putting the dirt over the slab roof safely is one of the most hazardous phases of building this house. As I mentioned , the small tractor with a front bucket is one method of moving dirt. It is also the preferred method.

TRAILER-SPREAD

If you have a light, compact car like a Volkswagen, and a small utility trailer like the smallest open-bed ones you can rent from rental outfits, you are also ready to move dirt. Pull the empty trailer to the source of dirt. Load the trailer, but use common sense as to how full to fill the trailer. Then drive the car and trailer onto the roof only after you have spread at least 6″ to a foot of dirt.

Whichever method you use, don't drive tractor, car or anything else directly over the tar or plastic. The tar, as thick as you have on your roof, will literally never dry hard and any vehicle will slip and slide as if on frozen water. You can get around doing this by hand by backing the trailer to the edge, dumping the first load and spreading it roughly. Then back over that load and dump the next. Keep leap-frogging backward until you have enough dirt spread to drive forward and make a circle, thus cutting down on the time it takes you to make a single trip. Naturally if you are using a small tractor, you dump frontward and continue the same procedure. Just don't drive on plastic or tar.

CRANE-SPREAD

Still another method that is to be considered, depending on availability and cost to you, is the crane. It could be the same one that you probably used to dump concrete. Use the largest bucket available and fill it with dirt using your backhoe. The crane will then very accurately place the dirt. The only disadvantage is cost.

Fig. 12-2. This small tractor was used until 3 feet of dirt was in place.

Whatever method you choose to use, be careful when working close to the edge. This dirt is loose and will roll and pack. This can easily upset tractors.

FINAL GRADING

After five weeks, if the 3 feet of dirt and the supports are still in place, it will most likely be safe enough to take your backhoe on top of the structure. Remove the bucket from the back. This is easily done, so don't take a chance carrying all the extra weight of the bucket onto the roof. Only the scoop on the front will be required. Once you are finished running around up on the roof with your backhoe, then put the aft bucket on and finish up the rest of your grading. The reason you can't continue to use a small tractor or trailer is because that will only move or carry approximately a quarter yard of dirt, while a backhoe tractor will carry one-half to three-quarters of a yard at one trip. You'll be thankful for the extra hauling capacity when you start dumping dirt over the edge to build up a sloped wall, if that is your design. Don't drive on the roof unless you have complete control of the tractor! Next to driving a grading vehicle on the roof, the most dangerous thing you can do is operate a bulldozer or large tractor close to an unsupported wall while pushing soft dirt. The weight of the vehicle pressing down on soft dirt displaces it. This dirt must go somewhere and it will try to push the nearest block or concrete wall away.

As final grading progresses, don't be surprised if you have to haul dirt to your site before you're done. I thought I had enough dirt to literally cover half the neighborhood, only to find out I needed another 125 tons of dirt. The need for additional fill dirt all depends on the lay of your land, the amount of dirt you have excavated and the design. If it is necessary to bring in additional dirt, check every price in town. I found good, clean fill dirt, not top soil, ranged from $1 a ton to $5 a ton. So you can easily add up the potential savings. If good top soil is required, the price is much higher. Therefore, don't use top soil for fill, and don't use fill for top soil. Only a few inches are required to grow good grass.

BACKHOE REMOVAL

As the grading and filling phase of your subterranian home draws to a close, you will find that this backhoe I suggested you buy is no longer required. Once the rough driveway is in and all the major grading is complete, begin to look for a buyer for this heavy equipment. If you made a good buy and didn't misuse your tractor, you will probably make enough profit to buy a large riding lawn mower with a snow blade and a small trailer in addition to grass cutting equipment.

GRASS SEED

As final grading takes place by hand raking and shoveling, be sure to plant the best grass seed and fertilize it as recommended by the experts in your area. Even grass seed varies from locale to locale. In final grading check into the use of sod—pre-grown grass. I tried to use it on a steep grade, and after the first heavy rain, it came tumbling down under its own weight. I had better luck planting fast-growing grass seed with good grass seed than I did with sod. With every condition being different, use your own good judgment and ask people in your area for advice.

13

Utilities

By now you ought to be pleased with your construction ability. You have a solid concrete shell floor, roof and block walls, From here on to final completion it will get easier physically, but at times the mental pressure will begin to get to you. If the building inspectors haven't hassled you or some neighbors are not up in arms, you're very lucky. But I'm sure you have all those minor problems under control.

PLUMBING

So now to the subject of plumbing. This is one of the easiest trades to do yourself. Most local codes will allow the home owner to do his own plumbing with a special home-owner permit. The best thing about plumbing, both water and drain, is that it's safe for an amateur to work with. Common sense will tell that drains always run downhill. The codes say that a good drain drops ¼″ for every foot of length. Of course, this rate of drop is not always true, especially in drains which are 2″ in diameter or smaller. If you followed the codes, you should have 4″ cast iron drains under the concrete slab. This is a nationally accepted code. However, once above ground the codes are as different as day and night. Some local codes allow the use of polyvinyl-chloride pipes, commonly called PVC pipes. Some places require copper or steel. Other places allow common plastic to be used in certain drains. All I can tell you is to check your plumbing codes. They are usually fair and easy to meet, especially the drain phase of plumbing.

WATER PIPES

As you read the regulations covering portable water pipes, you will notice that if copper is required, it will be either ½″ or ¾″ in diameter depending on how many branch lines are involved. Regardless of the size required, this pipe is really called hard copper tubing and it is bought in lengths approximately 20 feet long. There is a special copper pipe cutter that is inexpensive and invaluable when doing your own plumbing. Buy one. Do not use a hacksaw. After the pipes are cut and ready to be soldered into a fitting, they must be cleaned with a wire brush or steel wool. Each part to be soldered must have a bright, shiny finish. Once these parts are shiny, they are treated with a soldering flux paste. This is an acid base paste, similar to toothpaste, that cuts any trace of oil or grease off of the copper. Solder will not stick to copper if it is coated with grease. This procedure must be followed to insure a water tight connection.

Plastic (PVC) pipe can be used in place of copper pipe in some instances. I suggest you use PVC for cold water pipes where possible. It is so much easier to work with than copper. PVC won't rust, and it's easy to cut and put together. You will most likely have to use quite a bit of copper pipe to meet the codes. Don't waste it because it is expensive.

Another important thing to remember is to always buy the best quality valves, pipe, fittings, faucets, etc., to install in places that are difficult to get at for repairs. It's one thing to have a leak in an open area that is easily accessible for repairs, but it is a nightmare to get to a leak in some obscure cubbyhole without wall demolition. Quality is equally important when getting the spigots for bath and kitchen. Cheap units will rust away before you know it, and they are difficult, if not impossible, to repair.

MAINTENANCE

I realize this isn't unique to underground houses. But one of my reasons for going underground was to eliminate as much year-round maintenance as possible. It doesn't make much sense to do away with exterior maintenance problems and build in interior problems. The same theory is even more true when buying your deep well pump and installing it. Buy a reputable brand submersible deep well pump. They are easy to hook up although it is a somewhat physically back-breaking job when lowering that heavy pump and plastic pipe 300 feet down a 6″ or 7″ diameter steel casing. The point is that once it's down, it should be good for 10 years or more. Follow instructions closely, double checking the plastic for leaks. Check out all phases of wiring before lowering the unit down into the well casing. Be extremely careful because pipe and electric lines slide down the steel casing.

Pull up all of this pipe with the pump hanging on the end. This time the line is full of water and the water weighs approximately 62 pounds per cubic foot. It won't take you long to realize how heavy this is going to be. It could easily exceed 400 pounds.

ELECTRIC WIRING

If you are careful, you won't waste your valuable time on something as unnecessary as frayed electric wires. Question your wholesale suppliers. They are usually very cooperative. If you have a little extra money, I would suggest letting a professional do this pump installation ritual. It's a gamble doing it yourself.

If you followed my earlier suggestion and installed plastic conduit in the roof slab, your electric wiring runs will be much easier than if you plan to run all wiring around walls. If you are doing your own wiring I assume you must have some knowledge of the subject or know someone who does. Most electric codes will allow the home owner to work on his own house but won't allow anyone to help. Some locales even have a test you must pass.

ELECTRICIANS

If you have the money and are low in electrical experience, I'd suggest you let a registered electrician do the wiring. Electricity is one of the most difficult and most dangerous of all the trades.

If you do decide to hire an electrician, check around and get one for a good price. If you do it yourself, go to the wholesale outlets for parts. They will sell to an individual in large quantities. You save a bundle over a small retail outlet.

WIRING

If you're doing the electrical work yourself and you now have all the materials, the runs are made just as if it were the basement of a conventional home. The underground idea does not present any problems with wiring. Wires are clamped to the block wall in straight runs and all turns are made in 90° if possible. When roughing-in receptacle and switch boxes, they are mounted to the block wall by nailing and gluing a firring strip (usually 1″ by 3″), alongside of a hole knocked carefully in the concrete block. Locate the web of the block before you start. Mount ears of the receptacle box to the wood. Feed wires per code and continue until all the wires are in place. Then go around and cement in all excess areas around the box. If codes allow and non-metalic receptacle boxes are available, try to use them. Even if a wire should short out against the sides of the box, it would not ground out. I used them and they worked great. Once all wire is in place, use a felt tip marker (black or blue) and identify each wall. You should identify them by name, indicating direction and spelling out any other information you feel might be valuable in the future. Print this information in letters approximately 2″ high. Then take a clear black and white photograph. This way you can go back years later and locate wiring if necessary for additions or any other reasons. Be sure to wait until you get all the firring strips in place before you photograph, because it's good to know where the strips are also. As a rule of thumb, take the photo just before the sheet rock goes up.

TELEPHONE WIRING

I consider the telephone an important utility and since it is a necessity of life, why not prewire your house for the telephone just as you did for the electric? Here is another area where you can save money. The receptacle boxes used for telephone outlets are the same as the receptacle boxes used for a standard electric outlet. The boxes are installed into the concrete block exactly as if they were to be used for electricity. You can make arrangements with the phone company to install these boxes yourself, saving approximately $10 an outlet. Once you install the boxes in the locations of your choice, the telephone company will come in and complete the wiring installation. As I said earlier, the telephone company will work with you in most cases. Even though I installed a phone outlet in every room, I only installed two phones with jack cords. You can unplug either phone and carry it to any room in the house without paying for extra telephones. Ask the phone company how you can install the boxes and save money.

GAS

Natural gas or LP gas is a utility, but I'm only mentioning the word to remind you, as I did earlier—DON'T USE ANY OF THE LIQUID FUELS OR VAPOR GASES. Stick to electricity for everything. Remember, you don't have windows to open in emergencies.

All utilities—plumbing, sewers, electric, telephone—are basically the same when building either an underground house or building any standard structure. The point to remember is that when all walls are block, it is obvious that you will have a problem moving or changing any utility once the final walls are in place.

By now you should be seeing the light at the end of that overworked tunnel. The nicest thing about reaching this stage is the convenience factor. For the first time you can plug saws into any receptacle for temporary use instead of hunting for those elusive extension cords. For the first time a light bulb is in a temporary receptacle, instead of a drop light. Also, water is now available for washing or mixing cement or whatever else is required. The bathrooms may even be hooked up by now. This is your first taste of convenience. Up until this point you were really roughing it. Believe me, I remember what it was like to reach this plateau.

14

Woodworking

It's a fact that approximately 80 per cent of an underground house is concrete. You might think that your woodworking ability won't be tested. How wrong you are! Your woodworking skills will be put to the test in the way of hanging doors, trim and baseboard molding. This is where a radial-arm saw will be very valuable. The fine trim of an underground house is very similar to that of a conventional house.

NAILS, NAILS, NAILS

However, before the trim phase comes to pass, you will have to pound over 3,000 nails into concrete. If you have never driven a nail into a concrete block, you don't know what you're missing. It takes approximately 50 times the force to drive a nail into concrete than into wood. These 3,000 nails will be used to attach 1″ × 3″ wood strips to the block wall. The 3″ strips should be evenly spaced. Whatever you do, don't leave spaces wider than 18″. If you do the sheet rock will flex or possibly break.

Before these wood strips are attached to the block wall, you should apply some type of adhesive designed for gripping wood to concrete. Readily available at any building supply store is adhesive in tubes like caulk. It is applied with a standard caulking gun. This adhesive is commonly called *panel cement*. Apply a bead of panel cement to each board in a similar way to applying toothpaste to your toothbrush. Once these boards are nailed up and the cement hardens, these strips will never fall.

Let me explain about the type of nail you should use. There are two basic types available. One is called a *masonry nail*. It is common to use this type when attaching wood to solid concrete, like a floor slab. The other type of nail available is called a *cut nail*. They are rectangular in section and from my personal experience, I found that these cut nails were easier to drive into the concrete block. Drive a nail approximately every 2 feet. A rule-of-thumb to go by is this: The nail should be ½″ longer than the board you are attaching. As you are hammering these boards up, you will notice a technique is required to get this arm-breaking job done. You will soon see that the nails go into the block very tightly. Then if you give the nail one last hit just to make sure it is tight, then all of a sudden it will pop loose. This is because of the vibration caused by the hammer hitting the surface of the board and not the nail.

As the first nail in a particular board is driven in tight, move along about 2 feet and drive another. Once the next nail is started into the wood, hold your free hand on the wood to dampen the vibration. Continue this procedure until you have all the firring strips in place.

These firring strips are, of course, necessary to hold the sheet rock or paneling on the wall. However, the 1″ air pocket created as the covering goes up serves as a great insulator against temperature change or humidity.

HANGING DOORS

This is a bad feature of finishing off an underground home. When building a conventional home, it is standard procedure to use 2″ × 4″ interior wall studding. Therefore, interior walls always turn out to be approximately 4½″ thick. Because of this dimensional consistency, some companies manufacture a prehung door assembly. This is exactly what the name infers. The door is already attached to the frame with hinges and all final trim. The builder just inserts this assembly in place and "presto," the entire doorway is complete. Simple, isn't it?

But if you built an underground house to the specifications I indicated, you now have all of your interior walls at least 8″ thick. Add it up. The walls are 6″ concrete block—actually 5½″ wide. The firring strips on both sides of the wall are ¾″ thick each. And don't forget the two pieces of ½″ sheet rock. This adds up to a thickness of 8″.

Since underground homes have not really taken over the building world, no manufacturer makes prehung doors to fit 8″ thick walls. By now you have guessed it. An underground home builder has to completely build his door assembly from blank lumber, even to the point of drilling holes for and attaching door knobs.

This may not sound like a big deal, but it is. Hanging a prehung door assembly takes about half an hour. Building a doorway assembly from scratch takes even an expert carpenter 10 times as long to finish. Also, the quality is never as good as the factory-assembled units. Another fact you have to face is that you can't possibly build a door assembly as cheap as you could buy one if it were available.

There is a special tool available to insure that the door knob hole is in the correct position every time. This drill assembly is expensive and hard to find, but you must use it to insure accuracy. See if you can rent or borrow this tool. Buy one only as a last resort, because you will probably never use it again.

Adding Trim and Fixtures

While reading this book, you've noticed I'm giving my personal advice and opinions. Remember, I've already lived in my underground house for a year so I know the reaction your friends, the general public, neighbors and the inspectors will have. As for technical advice, any engineer can figure the strength of material, electric requirements and other specific sciences. It's the non-exact areas that can make or break the success of your home.

DECORATION AND LAYOUT

As you begin to plan interior decoration, do it with an open mind. I suggest you use unusual materials and designs and layouts (Figs. 15-1 through 15-3). Don't forget an underground home is unusual and it takes a special personality to own, build or design one. Once you are committed to going underground, you have to face the criticism anyway so you might as well add a few strange colors and designs. Why not have mirrors on the ceiling? Use whatever ideas you have hidden away in the back of your mind that you thought you could never use.

Fig. 15-1. Be original with your room layout.

BEDROOMS

We gave an open, warm feeling to each of our bedrooms by the use of special wall murals (Figs. 15-4 through 15-6). These murals are available from major wallpaper suppliers and range in cost from $40 to $150. Surely you'll find a design to fit

Fig. 15-2. Use unusual materials and designs in your room layout.

Fig. 15-3. These authentic, bleached barn boards are over 100 years old.

your personality and taste. The mural, such as the woods scene in Fig. 15-7, gives a warm sensation to any room, especially one without windows.

KITCHEN

Due to the fact that you don't have windows on any wall, you can really use the unobstructed walls to your advantage. In addition to murals, consider full wall bookcases or room-length shelves. You'll be surprised at the uses you will find for all of the additional wall space. Figures 15-8 through 15-10 are pictures of our kitchen. We used earth colors. Otherwise, it's very conventional even if it is 6 feet underground.

RECYCLED MATERIALS

Since most underground homes are built with energy savings in mind, why not carry that theme a little farther? Recycle—especially building materials. Authentic old barn wood is a good place to start. It will take some searching of the old farms and countryside to find wood that the owner will let you buy at a reasonable cost, or maybe even give to you. The end result of disinfected, trimmed barn boards in a hallway or family room is a very attractive appearance, complimentary to

Fig. 15-4. Cartoon-type mural on two walls in bright colors.

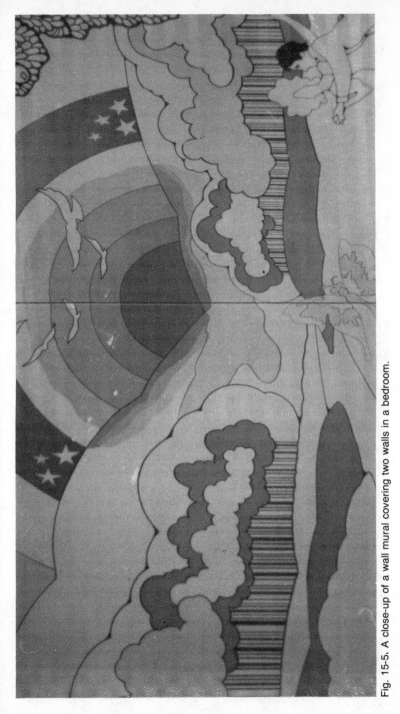

Fig. 15-5. A close-up of a wall mural covering two walls in a bedroom.

Fig. 15-6. This space-age wall mural gives a wide open feeling to this bedroom.

Fig. 15-7. Wall murals offer a room an open and warm atmosphere.

any underground house (Fig. 15-11). Also, you can find brick from an old building for steps, walls, shelves, floors, etc. Old brick or stone is relatively easy to find (Fig. 15-12).

FURNITURE

Then there's the furniture. Why not refinish old or previously owned furniture to suit your taste? You'll be mentally

Fig. 15-8. The kitchen of an underground house.

Fig. 15-9. This kitchen resembles many conventional homes' kitchens.

and financially rewarded if you take the time to salvage some of the past to go with your home of the future.

CARPET AND PADDING

Once you are ready for carpet and padding, shop around. Prices and quality are as different as day and night. As you probably realize, the carpet will be the single most eye-catching item in your house. Carpet just jumps out at you as you enter a home, especially if the quality is good. Also remember that the bright, light colors will reflect light and give a warm, dry feeling to your house. There is no use adding to the stigma of a cave by using drab colors. They will only add to the cave feeling some of your visitors may visualize.

PAINT

The same goes for paint. Always use light, bright colors. I suggest earthy colors: greens, yellows, browns, etc. Once again, the choice is yours.

Another suggestion I think you will find helpful when painting is the use of "sandpaint". This is a commercial product that is nothing more than a sand product added to laytex paint.

Fig. 15-10. This kitchen is totally electric.

Fig. 15-11. Old barn wood adorns the wall of this hallway.

It is thick, like paste, and is applied with a sponge. It really gives a different effect to sheet rock walls. In addition, you do not have to sand the sheet rock as smooth when sand paint is going to be used. Check it out. You'll like it. One disadvantage of this paint is that it can't be washed. So please don't use this rough surface where dirt will easily accumulate.

INTERIOR GARDEN

One of the spots in your house that will receive the most attention, simply because no one else has one, is your interior

Fig. 15-12. Brick used on this wall is over 125 years old. The stove also serves as this underground home's only means of heat.

Fig. 15-13. Hanging plants form the ceiling overhead in the garden.

garden. If you didn't include one, you'll soon wish you had (Fig. 15-13). In our garden area we used 2″ × 6″ wood blocks about 12″ long marking a walkway with an unusual pattern. We also hang many live plants over our head from the dome (Fig. 15-14). At times they even form a ceiling of live plants overhead as we walk through. Clinging ivy is another way to cover the block walls in your atrium.

Fig. 15-14. Plants hang from the skylight dome.

LIGHTING

Another subject that is only as limited as your imagination is lighting, especially in the garden. Give thought to dimmer switches, ground-level lights, hanging or side-mounted fixtures. Regardless of where, how or the quantity of light

Fig. 15-15. Your eating area will provide a place to explore endless decorating ideas.

Fig. 15-16. Light fixtures and carpet accent the bathroom.

fixtures, why not experiment with colored bulbs, especially blues and greens. Any good electrical supply store will have an ample supply to choose from. Once you start installing lighting outside, why not consider lighting the edges of the driveway with small D.C. exterior lights? Remember, the driveway is one of the few things that is visible from the road.

One fixture or device that we made use of was the commercial electric timers that are available everywhere. For

Fig. 15-17. Your bathroom is even open for a unique design.

example, we are experimenting with turning lights on and off in the garden as night falls and morning comes in. Another use for these timers is to turn your hot water heater on and off at times when hot water is not required. I realize lighting and timers are not restricted to underground homes, but they do give you another dimension to work with, especially for electrical conservation.

IDEAS

My wife collected pictures of interesting ideas of conventional homes from home and garden magazines. All we did was ignore the windows and then we found that many of the ideas could be used underground as well as above ground. If you don't have a collection of pictures, remember the local library has back issues of most magazines. For a few more ideas, see Figs. 15-15 through 15-17.

Air Handling

This is a phase that can cost you a fortune if you don't approach it with a great deal of logical thought. Due to the fact that the total air tightness will be underground, many of the conventional rules of thumb will have to be ignored.

HEATING

Any heating-ventilation contractor can install conventional sheet metal duct systems, such as the ones you will find in any conventional house. If you note any house with forced hot air, it has at least two vents per room. One is a return and the others are for incoming hot air. These vents are connected to a sheet metal duct system that comes from and returns to a central heating plant. Other possible ways of heating include electric and hot water baseboard heat. Many people use only woodstoves to heat their underground home. The metal duct system is by far the most expensive, most difficult and most inefficient system to use in an underground house. The reason is that the contractor would have to break through concrete walls at least three times per room. In my case, with ten heated rooms, that would be at least 30 holes broken by hand in concrete block or solid concrete. This costs time and money, not to mention weakening the structure.

A second negative reason not to use a metal duct system is that the sheet rock walls would have to be built around the duct to cover it when finishing a room's interior. This in itself is a time consuming job.

My system is simple, but efficient. It cost approximately one-tenth of the cost of a conventional hot air system and in many ways it is twice as efficient and flexible. The general idea that I used follows.

Start with the room where your woodstove is located. This should be in a centrally located room. As this room heats up by the radiant heat of the stove, it is nothing more than a big hot air plenum, serving the same purpose as a plenum chamber or a furnace. By installing a miniature motor approximately one-hundredth of a horsepower with approximately an 8" diameter fan blade into the wall, the fan will draw hot air out of the main room into the next room. By using a variable speed control such as a light dimmer switch, you can vary the speed of this fan motor. Thus you control the volume of air that flows through the wall. By repeating this procedure in any adjacent room, you can circulate the air as necessary.

Do not forget that by drawing air into a room, there must be a vent to allow the same volume of air to escape. In some cases, I used a small grill for this escaping air, but in most rooms the doors were cut to clear the carpet by approximately ½". This ½" clearance allows the pressure to escape very satisfactorily. Of course, if the door is open, there is no problem at all. The air flow will continue to circulate by using only this small in-the-wall motor and no duct system. The only additional venting necessary is to have a clear unobstructed path for fresh, exterior air to be drawn in as the hot air goes up the chimney. In my house, as in many others, all that is required is to vent air into the atrium or garden area. If your garden dome is not air tight, as most are not, further venting in the dome is unnecessary. This is the main reason for not making this foundation a solid unit.

COOLING

In the warmer months when the wood stove is not used, the fans serve the same purpose of keeping the air moving, thus avoiding a stale or musty odor. By using a standard, good quality room dehumidifier in one or more rooms depending on the severity of the humidity problem, and letting these fans circulate the room air past these dehumidifiers, it is relatively simple to control the humidity in the entire house without a complex system of duct work or expensive central heating-humidifying unit. The disadvantage of putting a hole in the wall is that now sound can travel very wasily from room to room. This condition can be nearly eliminated by side-stepping this air flow with a simple duct system you construct yourself as the wall board goes up.

This allows the air to go through the wall, then travel behind the sheet rock until the grill vent is installed on the opposite wall approximately 4 feet away from the concrete opening. This method is only a minor adaptation to the direct air flow, but it is a major barrier to sound waves.

In my house these fans run nearly 75 per cent of the time, year round. This may sound like we're using a great deal of power, but don't forget that each one is only one-hundredth of a horsepower or a collective total of one-fifth of a horsepower for my entire house. Most conventional furnaces have motors that draw more than twice the power that these require. This is just another advantage of building underground.

17

Domes and Skylights

From the data I have collected on underground homes, I figure that 85 per cent of them have either a skylight or a dome. Approximately 75 per cent have some type of interior garden or atrium, or courtyard. I highly recommend that you consider including a garden in your plans due to the absence of windows. You need a method of natural light access to eliminate any trace of a cave-like atmosphere. Because your house is below grade level, a dome is much easier to build and install than if you tried to put one on a conventional house roof that is 20 feet off the ground.

INDOOR GARDEN

My garden is approximately 8 feet wide by 16 feet long. This seems to be the average size, especially when you consider the cost of building or buying a dome. The dome or skylight is another place where a person handy with tools will save money. I built my dome in the shape of a pyramid and covered it with acrylic plastic ¼″ thick. It measures 16 feet wide by 24 feet long by 8 feet high. To see my dome under construction, refer to Fig. 17-1. Figure 17-2 shows the method of construction.

As for building the super structure of a dome, all you need is a hand circular saw with a special aluminum cutting blade, a square, a tape measure, a good aluminum welder and a place to work. Don't use any material for structural support but aluminum, because it's light weight, easy to cut and it won't rust or need painting. You can buy aluminum from any steel supplier listed in the yellow pages. Check at least three or four suppliers because the price of aluminum varies quite a bit. Aluminum is sold by the pound, so the more you buy, the cheaper the pound rate. Therefore, buy all you need at one time and pick it up if possible. Shipping of irregular shapes is always costly. If you decide to build your own, try to stick to a shape similar to my pyramid style shown near completion in Figure 17-3. Avoid curves because the bending of plastic covering is very difficult without the proper specialized plastic heating equipment.

COST OF DOME

If you are still undecided about whether to build your own dome or buy one already complete, consider these facts: I completed my dome (technically not a dome) at a cost less than $3,000. The equivalent structure commercially built would have cost more than $9,000, not delivered and not totally assembled. I was afraid to ask about a delivery charge. There is another drawback—you'll find only a few dome manufacturers interested in talking to a private home owner since most of their work is with commercial building contractors specializing in shopping malls and office buildings. However, if by now you are considering doing without a dome or a skylight, I'll try again to convince you to include one at any cost.

Besides the element of natural light, consider the pleasures of a year-round growing area, for exotic plants or a small vegetable garden.

In addition, an atrium will give you the ideal place for natural ventilation or air-draw because of its height above grade level.

Fig. 17-1. Dome under construction.

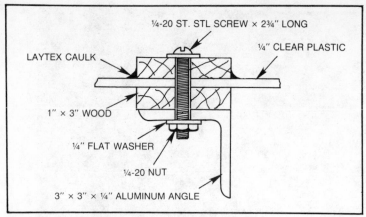

Fig. 17-2. Typical section of dome structure.

Now for the major reason. A garden or atrium covered by a dome of some type will give you a place to provide emergency exits in order to comply with some local building codes. Check codes carefully as you design your garden and dome. As I've said in almost every chapter, consider all possibilities before locking in on a design. See Figs. 17-4 and 17-5 photos of my garden in mid-construction, and Figs. 17-6 and 17-7 for photos of it near completion.

SKYLIGHTS

Skylights, as opposed to domes, are usually considered only to let light in. They are mostly transluscent, and probably will be much easier to buy than build because the shipping costs aren't prohibitive. Check the large building supply stores as a source. These are a few of the possibilities. It's your choice once again.

CONSTRUCTING YOUR DOME

If you decide you want a dome, find a good aluminum welder who will work with you. If he can't work on your site, then at least be near enough to make moving the welded frame possible (Fig. 17-8). Never install the plastic panes until the dome is in place and set on a stable foundation. The flexing will break the plastic panes every time. Check prices and explain

Fig. 17-3. Dome near structural completion.

Fig. 17-4. View of the garden during mid-construction.

Fig. 17-5. Progress is being made on this dome.

Fig. 17-6. Dome is near completion.

Fig. 17-7. Dome in place over garden.

your project to the welder. Keep looking until you find one who is interested in your underground house, not just your cash. Try to arrange to do the sawing, fitting, holding and grinding yourself. Let the welder weld. You'll both be satisfied and the finished product will show it. If in doubt about strength, return to the engineer who helped with the roof. If he was accurate with the roof strength, the dome will be a breeze.

Once you find this person capable of welding aluminum and he agrees to help you, you have to have all aluminum angle and flat bar delivered. Aluminum angle of 3″ × 3″ × ¼″ and aluminum flat bar of ¼″ × 3″ should be sufficient to build a normal dome. To avoid cutting aluminum at the wrong angle or too short, use cardboard to make a template at the joints. Use a good tape measure and triple check the dimensions after tack welding and before permanent welding. Continue this procedure until the dome's super structure is complete. Once welding is finished, use a grinder to smooth welded joints so that the plastic and wood will lay flat.

Once your framework is complete and ready for installation, you need to construct a foundation to set this structure on. I stacked 8″ concrete blocks from roof slab to ground level (Fig. 17-9). Then I set 8″ by 9″ railroad ties around the perimeter. Finally the structure rested on the ties and was bolted down. The reason I stacked block with no mortar was to let the dome breathe and to allow some moisture to seep through to the earth inside the dome.

In all fairness to you, beware of this method. I understand that it may be against some building codes in some locales. Let me mention something very interesting about the indoor gardens of underground homes. One inspector will say that a garden area like mine is an outdoor area and that exterior building codes apply to electrical outlets, water lines, etc. The next inspector will say that the same area is definitely an interior room and that interior codes apply. Let them fight it out. I consider it an indoor area.

Fig. 17-8. Method of moving dome after construction.

189

Back to the completion of this dome. There are only a couple of choices of material for panes. There are many acrylics on the market suitable for panes in a dome. One such example is plexiglass. Check all suppliers for prices of 4′ × 8′ sheets of ¼″ clear acrylics. Do not use anything more than ¼″ thick. The expansion and contraction is a factor to consider when using as much plastic as you will use. One-quarter inch will be strong enough, yet the weight will not be too great.

You also have the option of using tinted colors in acrylic plastic, but they are more expensive. But do give some thought to the idea of installing tinted panes. You have to live with it for a long time. However, I personally suggest you stick to clear because the sky and clouds will begin to look strange through pink or yellow or whatever tinted panes you might choose.

Another thing I would suggest while designing your dome is that you do not make any of the sections bigger than can be covered by a 4′ × 8′ sheets of acrylic. Buy these by the gross from a wholesaler. Don't buy them from the local hardware store, or you'll go broke. Space the bolting pattern so that a bolt ends up within 3″ of each end of the pane and at 12″ intervals. There is an easy way to make sure that each sheet of plastic is cut to exact shape and you don't make a mistake and ruin a costly 4′ × 8′ sheet. With the help of two extra people to help hold a sheet in place over the frame, move it around until it fits an open space with a minimum amount of waste (Fig. 17-10). Now use a straightedge and a crayon to draw a straight line where you want to cut. Continue this process until all panes are cut. As you cut a pane, put it in place to avoid scratching or breaking. Whatever you do, leave the paper on and keep the sheets out of the sunlight until cut, then remove paper at installation. If the sheets are exposed to sunlight for more than a few days, the paper becomes brittle and is difficult, if not impossible, to remove. Ask you supplier. He may be an expert on plastic or just a part-time salesman, but it won't hurt to ask. After all these procedures are followed and the dome is built, installed and all the panes are in place, the

Fig. 17-9. Stacking 8″ block forms a mortarless foundation.

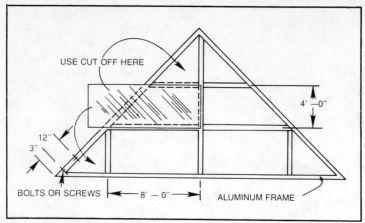

Fig. 17-10. Method of cutting plastic to correct shape.

final step is to caulk each seam using an acrylic caulk. Any good quality will work fine. Finally, the dome must be bolted to the railroad ties by using lag bolts, approximately 6″ long.

As a safety reminder, do not watch the welder as he is welding. The light flash will burn your eyes. Even a reflection off of a wall will cause a severe burn or possible blindness. Be very careful if welding is a subject you are not familiar with.

Another possibility for building a dome is using prefabricated panels in triangular shapes. Look in the classified section of the how-to-do; plus magazines, such as *Mother Earth News, Mechanics Illustrated, Popular Science,* etc., for geodesic panels.

18

Underground Home Ceiling

Have you ever stopped to think recently about what your ceiling is going to look like when your house is complete? I realize that with all the other problems of building an underground home, you haven't really had time to worry about the cosmetic effects of the concrete slab—your roof. However, if you stop and think about it, you'll discover this 10″ concrete slab has two sides, one of which is on the inside of your house. This could possibly be your finished interior ceiling.

ROOF SLAB

If you poured your roof slab in one piece over a sheet of plastic and you have removed all the scaffolding, you will notice the unique pattern formed in the concrete by the plastic. Most likely, this will be suitable for a finished ceiling texture. All that was required was to paint the concrete roof with flat latex paint.

I've mentioned this natural way of completing your ceiling, because although there are other ways of finishing your ceiling interior, they are definitely a job to install.

A fact you must accept is that, with all of your shoring, reinforcing and other construction involved in pouring the roof, it is not logical to expect the underside of your roof to turn out smooth and level. It will definitely have high and low spots, thus preventing you from attaching sheet rock directly to the underside of the concrete.

The uneven surface is not the only determining factor. Probably the largest barrier you will discover is that you cannot drive concrete nails into your ceiling. There are two reasons for this. First, the concrete, if it has correct quality, is too hard and brittle to accept nails. The main reason is that it is physically difficult to hammer over your head for any length of time or apply enough hammering leverage.

There is only one way to attach sheet rock to the ceiling and it isn't easy. Here is the method that will work: Use a ¼″ drill bit, drill a hole approximately 1½″ deep and insert the appropriate part of an anchor-bolt assembly. These are available from almost any hardware supply store. I suggest you use 1″ × 3″ wood strips, 8 feet long for ease of handling. You should first place a bead of panel cement on each wood strip and then put anchor bolts every 2 feet for a typical wood strip installation. Now all you have to do is to continue this procedure throughout the house in each room you want sheet rocked.

If you don't like either of these methods of finishing your ceiling, there is yet another alternative to ceiling decoration.

CREATIVE CEILINGS

This alternative is the method I have on the ceiling of two of my rooms. All that is involved is a little patience, originality and time. The pattern is nothing more than pieces of wood cut at random lengths, widths and thicknesses. The largest piece is approximately 8″ long. The reason for the short lengths is that panel cement is the only thing holding them in place. No bolts, screws or nails, only glue. Most likely the smaller pieces will stick without support until the glue sets, especially if you tap it gently with a hammer to force a tight seal. Experimentation on your part will soon develop a style and approach that will make a super ceiling with a personal touch. An additional value of this method is the fact that wood is an excellent insulator and a little extra insulation never hurt anything.

As in previous chapters, I have tried to be honest with you on the pros and cons of each step as I see them. The negative side of this method is the extra wood is looked upon as additional fire hazards by the inspectors. This might cause some contention, so consider this negative possibility.

If you do decide to glue something to your ceiling, remember that almost anything can be glued up if you use a pole brace until the glue dries. The glues on the market used for construction are very strong.

In short summary of ceilings, if you plan on having a conventionally smooth sheet rock ceiling, extra effort will be involved in securing the sheet rock and getting it on a level plane. The easiest and most artistic approach would be to glue something directly onto the concrete or apply a stucco or paint of some type. Keep your imagination working overtime!

19

After Basic Completion

Once the basic construction part of your underground home is complete, you can and should relax for a while before going on to the fine details of finishing.

FINISHING TOUCHES

These details include driveway surfacing other than basic gravel, landscaping other than grass and standing trees and so forth. On the inside these finishing touches may include the shop work bench you've always wanted, or storage shelves everywhere. These are all things that are nice to have and you want them, but they are not essential to a final inspection by the bank or building codes department. As a matter of fact, anything you build is subject to inspection. Therefore, the less there is to inspect, the less chance you have of hitting a snag. Take your time adding these finishing touches.

STORAGE SHEDS

Another subject I want to cover that is particularly pertinent to the underground house is the ever popular outside storage shed. These are the metal sheds that often rust two weeks after erection. It never ceases to amaze me that people will buy a $90,000, 3,006 square foot home with a three-car

Fig. 19-1. Driving up to an underground house.

Fig. 19-2. This long driveway leads to an underground house.

Fig. 19-3. Entrances to an underground house.

Fig. 19-4. An underground home.

garage and full basement, and within six weeks of moving in they find they need outside storage. So they buy one of these storage sheds for a couple of hundred dollars and the tax assessor sees it and up go the property taxes. They continue to pay a premium for having the shed year after year, even after it's rusted beyond recognition. What I'm leading to is this: These sheds seem to be an intricate part of middle-class American homes, probably because almost every neighbor has one. Somehow I can't imagine building an underground house, spending countless hours getting grass to grow on the roof, side slopes, creating that back-to-nature attitude and then methodically plopping an outbuilding of any type nearby. It's not a logical move and is bound to be an eyesore even if it's well constructed. You must remember that your structure is not seen by your neighbors, so the outside shed, garage or outbuilding is the only monument visible to the outside world. Can't you just picture driving up to your underground house (Figs. 19-1 and 19-2) and seeing nothing except a rusty storage shed in the middle of three acres? No way! If your house can be seen from the highway, why make the world look at your storage shed? I'd even go as far as to request that you keep the lawn trimmed and neat. Let's not give opponents of underground homes anything to pick at. We may be an endangered species.

THE ENTRANCE

Another thought I'd like to pass on is why not give your entrance (Fig. 19-3) a distinctive appearance? Maybe an unusual mail box, or just a nice display of evergreen shrubbery. Unless there is a sudden surge of popularity in building underground, your house will definitely be a novelty and the curiosity seekers will drive by to take a look, even though they may not see much from the outside. They will form a lasting opinion of your home by what they see from the road. See Fig. 19-4 for a view of my home.

20

Getting Used to Living Underground

Building underground is one thing to accomplish, living underground successfully is another. You must realize, as you are building, that there is no way to see if you will like living underground until you actually are. Spending a few hours touring and visiting an underground home doesn't give a true indication. Hundreds of people have been inside my home for a short period of time, but they still don't know for sure what it's like living there. The sad thing about this point is that it costs a great deal of money to find out. If you don't like what you've built, there is a big question as to its value. If you're lucky, you could sell it at a profit, but on the other hand, your potential buyers are few and far between.

NO WINDOWS

Probably the first thing you adapt to easily, as we did, was the absence of windows. You've probably always had a window to look out of or to open. Here is where your dome and indoor garden will save you.

STRANGE SENSATIONS

For the first few months of living in our home we experienced an odd feeling caused by the absense of noise. In a conventional house when you turn the television off late at night and the children are asleep, if you sit still and listen, you might say things are deathly silent. Well they are in comparison to normal activity. The average person would go bananas in an absolutely soundless room.

An underground house is somewhere in between these two extremes due to the fact that these block walls absorb the sounds that most people never realize they hear. Noises such as hot water heaters warming up, or the presence of car noises on the nearest road, or airplanes, or dogs barking are all considered "normal." These are now almost totally eliminated. The strange effect is this: If your daily routine takes you to the outside world, such as shopping or school, your ears become tuned into the sound level around you and your speech level is comparatively higher. However, when you walk into your underground home especially for the first few months, you will find yourself talking louder than necessary by force of the habit from trying to be heard normally on the outside. The absence of background noise isn't noticed and you continue talking as if it were present. This is something you get used to in a short time so it's nothing to get excited about. Also I should add that the layout and configuration of the rooms makes quite a difference as well as the carpet quality and the amount of glass.

RADIO AND TELEVISION RECEPTION

This is something that never crossed my mind as I designed my house. Once we moved in and turned the radios and televisions on, we immediately noticed that in some rooms absolutely nothing could be picked up on the radio. The waves could not penetrate the earth and concrete. Of course, as you get closer to the doors or your dome, if you have one, the television and radio waves will filter in giving some reception. Even close to an opening, the reception leaves much to be desired.

Our solution was to use the aluminum structure of the dome as an aerial for both television and radio. The amount of aluminum available and its configuration make quite a difference. So you may have to get a service man good with television and aerials to stop by and show you the best way to attach and run the wiring.

Bad reception is easily overcome if you run a master aerial lead to each room as you are finishing the walls. This wire is not covered by codes and is harmless to work with.

CABLE TELEVISION

If you live in an area where cable television is available, I would suggest you check into the service to insure good reception, especially if you are an avid television fan.

As you see, things are somewhat of a new experience and quite exciting at times.

21

Alternate Sources of Energy

Alternate energy, now there's a catch-all phrase! Since I began building this underground house I have heard more stories about alternate sources of energy than you would believe.

My definition of alternate energy is any energy source for which you don't receive a bill in the mail. Of course, energy is synonymous with utilities when discussing home use. The two basic requirements are heat and electricity.

HEAT

Some people who have heard unfounded stories about underground homes not needing a heat source are definitely of the wrong impression. Heat most certainly is required. The good thing about building underground is that it only takes about 25 per cent of the heat required for a conventional home. As you can figure, it doesn't take long for a 75 per cent annual fuel bill savings to add up to a substantial amount. Then there's the obvious fact that sometime in the near future fossil fuels, as we know them, will be unavailable to heat your house at any cost. So you can look at the underground idea as doing your part to save energy. Now that I have convinced you that some small heat source will be required, I'll help put into true perspective the alternate methods of obtaining this form of energy.

WOOD STOVES

I have come to discover that this back-to-basic piece of equipment called a wood stove is probably used in over 50 per cent of all underground homes. I base this fact on information given me by other underground home builders that have contacted me. I've taken no scientific survey, but I'm convinced that wood stoves are the most popular heat sources in underground homes.

Like any other commercial product on the market, there are stoves of excellent quality and stoves that are outright dangerous to use. You will only get what you pay for in a wood stove: so be prepared to buy "quality." Most of the brands are sturdily built and have excellent air flow. You will find quite a few manufacturers of wood stoves that distribute regionally. Therefore, they are not available to everyone, everywhere. There are good books written on wood stoves. So I suggest you make up your mind intelligently by reading up on the subject and by physically examining as many brands as possible until you're satisfied that you know what you are buying.

Here is a good example as to why a good stove is important. There are many scientific results available from different manufacturers indicating that one brand burns twice the wood of another brand under similar conditions supplying the same heat. This is a fact. So you see you could end up cutting twice the wood necessary to get through a winter.

As for my underground home, I don't have to buy wood. But I do realize that wood could be a costly fuel if you have to buy cords of wood year in and year out. But even if you paid premium prices, it would still be cheaper than the oil-base fuels. Besides, wood is a replenishable natural resource. If you have to buy wood by the cord over any extended period of time, you're doing something wrong. Wood is everywhere. Some builders will even pay you to clean up the scrap wood after a new house is built. Many people will pay to have a piece of land cleared of the trees, and you keep the wood. All you need is a good chain saw, energy, and the old station wagon mentioned earlier. I won't dwell on this subject since it is only remotely related to underground home building. All that I'm saying is, don't buy wood, just look around because it's everywhere for free.

COAL

Just as I use wood as a fuel, some people use coal. In some sections of the country it is easier and cheaper to obtain than wood. There are problems with the odor and the mess, but it's cheaper and easier to handle than wood. So check the price of coal as compared to the availability of wood.

The important thing to remember if you use a stove is to locate your heating unit centrally. Your chimney is the main part of your natural heating-ventilation system. Put it near the center of the living area. As your hot air rises up the chimney, it draws air from the point of least resistance, and that is out of the living area as long as you provide a method for fresh air to be drawn inside the house.

In an underground house the hot air rises up the chimney. This draws air out of the adjacent room creating a slight vacuum. This vacuum is filled by fresh, colder air from the outside, thus creating a continuous flow of air naturally. This idea of air flow may need mechanical assistance such as a small fan or duct system. The exact size and layout of your home will determine this. You may also use vents or lowered panels in doors to create a flow of fresh air. This pattern of air flow will continue as long as the air temperature is warmer inside than outside or a slight breeze is blowing past the chimney.

Of course your building inspector will ask, "What happens when the wind doesn't blow and it's 100° outside and you have a fire in your stove?" He is technically correct, although not practical. Since this condition will only come up in the summer months, it is safe to say that your doors won't be continually shut for an extended period of time (a week or more). But in case your inspector requires additional assurance of ventilation you can install a small fan and vent from the exterior, or you can draw fresh air from the indoor garden which I hope you included.

This air circulation is a subject that your inspector may harp on or not even mention. It depends on your local inspection system. If it's a big deal to your local officials, rely on your engineer. You probably know him pretty well by now.

For your own information, let me tell you that you can take an underground house the size of mine (2800 square feet of living area), close all exterior doors, burn your stove for at least five days while living normally inside and the atmosphere inside will hardly be stale. The fact that an underground home is more airtight than a conventional home scares some people into thinking they are going to lack fresh air. It just isn't so. Normal traffic opens and shuts exterior doors numerous times a day. I do suggest that a vent of some type be installed on the exposed exterior wall, and that it feed directly to the interior only as a safety precaution. This is a subject that your local code will control probably in great detail. If you have a vented dome, you're even better off.

SOLAR HEAT

Solar heat is another catch-all phrase. Breaking it down, it means any heat from sunshine. This includes a very simple system of simply opening curtains as the sun shines and closing them as the sun sets, or it could be the complex systems of piping, valves and pumps used on conventional houses adapted for solar heat.

It is my opinion from my research that complex solar heat systems will cost much more to build, install and maintain than you could ever retrieve in heat bill savings in a lifetime. Remember, I'm going on the assumption that your heat requirement will be one-fourth that of a conventional home.

I do encourage you to use solar heat and to do all possible investigation if cost is not a major factor. Solar heating is definitely a subject whose time has come, and everyone should support its use and research. The only drawback is the initial cost. Another negative feature about solar is that you lose the homey atmosphere created by a fireplace.

SOLAR ELECTRIC

Please do not be confused with the two terms, solar heat and solar electric. Solar heat is here today. Solar electric cells for domestic use are in the future somewhere, probably five or six years away. Or it's possible a major breakthrough will never be developed in the production of solar electric cells (also photovoltaic cells). The space industry has put these cells to use, but that is about the extent of their availability.

A WORKABLE ELECTRIC SOURCE

There is, however, a source of electrical power that is less expensive than the power company. You should be well versed on electricity and mechanically inclined to pursue this method. I'll describe the basic idea and you can take it from there since it is only a take-off of the philosophy of underground homes by way of encouraging self-sufficiency.

The alternative electrical system is this: With a small diesel engine, such as a four-cylinder Volvo for example, you can couple this with a direct current (D.C.) generator that charges heavy duty storage batteries similar to the ones used in commercial equipment. By a series of regulators and switches you can keep these batteries fully charged with a minimum use of diesel-fuel. The catch is that you will have to wire your house for D.C. use instead of the usual A.C. from the power company. Most of your light bulbs and small appliances will oeprate sufficiently off of D.C. current. Only the big electric appliances require A.C. This system would be somewhat expensive to set up, but it would pay for itself in a short time, especially the way electric costs have gone up recently.

The slight inconvenience might be worth the advantages gained.

WIND POWER

What about a wind generator? A really interesting source of electrical energy that is free is a wind generator. There are approximately 10 good manufacturers in this country. Each claims to be the most efficient. All of them work basically the same. Some use storage D.C. batteries, others use inverters without D.C. batteries and provide A.C. current. One thing to be sure of is that your area has an average wind speed necessary to turn a windmill. Since the wind blows at different speeds from day to day and each wind generator's conditions are different, it's hard to be specific about what conditions are required. From my research I have drawn the conclusion that if you had 15 miles per hour winds 50 per cent of the time, using batteries and a 5-foot diameter blade on the appropriate generator, you could probably do away with the electric company. Call the local airport for basic wind information.

22

Underground Home Publicity

There is a fact that you must face when getting involved in any project as big as building an underground house. As with most unusual projects, many people will be interested. It seems that with the energy shortage, underground homes are surely a subject to get attention. Of course, there are many types of attention that you can get—negative, positive, private and public.

POSITIVE ATTENTION

Fortunately for you as the home builder, the people that give you publicity, whether it be by their request or yours, are fairly easy to categorize. As you read on, you will see what I mean.

I have found one fact to be 99 per cent true. The groups of individuals who personally ask to see your house are almost always friendly and not likely to cause you any problem. You very rarely will have a person ask to see your house and then bad mouth it behind your back. If you build your underground house correctly, visitors will go away impressed. They will also become your best moral supporters. You will be surprised by the number of friendships you make that will continue after the house building has been completed. This is reason enough to build an underground house—we all need all the friends we can get.

NEGATIVE ATTENTION

Just as the private citizens who ask to see your house are 99 per cent friendly, you can rest assured that 99 per cent of the public or regulatory personnel who see your house will have a negative opinion. In all fairness to these people, they usually form their opinions through the eyes of their specific jobs—zoning, health, fire, insurance or building inspectors. Don't be surprised if the dog catcher even gets into the act! Let me explain what I mean by seeing an underground house through their job titles as opposed to their personal interest.

In the course of my complicated maneuverings with the county and state officials, I had one inspector who gave me real problems as he acted in the manner and capacity of an inspector. As a matter of fact, he was downright uncooperative. However, a week after the inspection, the same inspector contacted me and wanted to know if he could show my house to his family. This time everyone was as friendly as could be. Once back on the job, he reverted to his old self.

PUBLIC ATTENTION

This is the most critical type of attention you will receive. It can make your adaption into the community smooth, or if the attention the newspaper, radio and television give you is negative, the neighborhood will be convinced that your house is a black spot in the community. At this point, I will put your mind at ease and tell you to relax. When a newspaper reporter or television station contact you for an interview, they are almost always forward-thinking, intelligent individuals who like to see people doing individualistic projects, especially saving energy. The time is right. These reporters can be your best allies if you get into any real hassles with officials. Reporters, by nature, will see that no one pulls the wool over anyone's eyes simply by constant exposure in the media.

I do, however, have some good advice for you as you prepare for your interviews with the media, be it television or newspapers. Know what you are talking about, don't make dumb, specific statements. First of all, they only want general information that the public can relate to. For example, if a reporter asks you how strong your concrete roof is, don't answer that it will hold up exactly 795 pounds per square foot. First of all, no one knows exactly how strong your concrete is and secondly, someone will begin to question your judgment. Whatever you do, don't pin yourself into a corner by broadcasting specific facts about your house that the public does not need to know.

All the reporters I have talked to have been friendly, so be sure to act accordingly. They are great people to have on your side to spread the word about the good points of your underground house.

The one reservation you must have around reporters is to be exact as to what you want to be *on the record* or *off the record*. This, of course, has nothing to do with underground homes, but since you probably are not familiar with interview procedures, I will only tell you that once a statement is made, it cannot be retracted. However, a reporter will not report anything preceded by the words, "This is off the record."

PRIVATE ATTENTION

This is the same as acceptance by or rejection by the neighborhood. The less negative attention you receive publicly, the less private attention you are likely to receive. Accept the fact that building an underground house is an attention-getting project. Use it to your advantage and do not let it cause you a problem.

23

One Year Later

My underground home has been completed for over a year now. For you, the reader and potential underground home builder, my problems will be your gain. In reading the previous chapters, you learned of my experiences dealing with day to day problems. This year has given me time to collect my thoughts and change my opinions on some things. Some of the things I thought to be insignificant were not, and some of the things I thought to be important and critical were no problem at all.

First, I'll go over a few items that gave me a problem or at least some concern during the past year.

DOME ALIGNMENT

As I located my house on the piece of property I had bought, I centered the structure with the front doors facing the only direction feasible. I thought nothing about this layout at the time because I thought if I was totally underground, what difference would it make how the house was situated. For the most part, there was no problem. The slight inconvenience I did discover was caused by the way the sunlight comes through the panes of the dome. As it turned out, I had two corners of my garden area under the dome. About 15 per cent of the total area was never directly hit by sunlight. This doesn't affect the garden in any way, except for losing a small growing area. Needless to say, a real professional could have figured the path of the sun and the shadow it would cast if he had given it the time and concern.

Another little discovery that I made, once I actually settled down to a daily routine in my house, was that the dome sweats year-round. Only in extremely cold temperatures do the droplets freeze to the panes of plastic. Otherwise, the condensation droplets accumulate and cling to the panes. This is fine, and they soon evaporate and dissipate into the air. However, all good things have at least one drawback. These droplets cling to the panes as high as 22 feet overhead. If the wind is blowing slightly, all is well. Only when the wind velocity nears 40 miles per hour in gusts do the panes vibrate, thus shaking the droplets loose to fall to the ground. Still, it is not a big problem unless you are walking through the garden in the early morning around sunrise when the most drastic temperature changes take place. Unfortunately, this is the time we are usually crossing through the garden from the bedroom to the kitchen for the first cup of coffee. If the time is right and the wind is blowing hard enough, you can get a nice shower. This only happens once in a while, so I consider it a unique feature of underground living. I'm not about to do any major construction to eliminate a few drops of water. This is the type of phenomenon that was not expected. Only a year's living experience brought it to light.

DOME FOUNDATION

While on the subject of the dome, I have still another fact to share after one year. The building codes call for any structure similar to my dome to be anchored solidly to the roof structure. If you recall, I did not do this. I stacked concrete blocks on the roof slab using no mortar. The reason for no mortar was to allow normal water seepage for the indoor garden soil. The inspectors said that this would cause a problem, and I was cited for noncompliance. This point has still not been resolved. The fact is that the stacked block works extremely well. Only a limited amount of rain water actually seeps through, and the cracks allow air to filter through into the top of the dome. The statement that the earth will settle against these stacked, unmortared block and cause them to cave in is unwarranted. My block are exactly in the same place I put them 18 months ago. One thing I did do after one year was raise the dome 8″ higher by placing another row of block around the perimeter. The reason for this was to allow a build-up of dirt to create more of a slope for rain to run off.

SUN HEAT

Another interesting fact I discovered about having an enclosed garden area is the true power of the sun to heat. Now, one year later, all of the armchair engineers are telling me, "I told you it would be hot in there." Of course anyone could have predicted that, but I don't believe anyone expected the temperature to get as high as 165°F, which it actually did. Do you know that 165°F literally fries tomatoes on the vine? This was a problem that I discovered only after I lost quite a few good yielding tomato and pepper plants. The solution to that problem is simple. If you are in the midst of a hot spell, spread a thin sheet of polyethylene (plastic) over the plants. This is the same plastic you used for waterproofing the walls and pouring concrete. It works well. Additionally, during the extreme hot days of August I used a portable electric household fan directed upward to circulate additional air. It is not practical to install a fan of this size permanently since it will only be used approximately 10 days a year. These are truly little bits of information that only time can reveal.

CRICKETS

Many people ask if I have an insect problem. No, I do not have an insect problem. Specifically, I have a cricket problem. Right! Crickets. Those black grasshopper-like creatures that supposedly make noise by rubbing their legs together. One year later, I can truthfully say that spiders, ants or bugs of any kind are literally non-existent. Not many normal, above-ground homes can say that.

I have been told that an indoor garden like the one I constructed is the perfect living quarters for crickets, and I have verified that fact. Now, before you get excited and conclude that my house is overrun with crickets, I'll be specific. First, they seem to congregate in the garden only. Occasionally one will venture into another room, but a normal house will have that same problem. The fact that they stay near the humid, garden-dome area makes them easy to control. Many of the commercial insect sprays do the job. It doesn't eliminate them completely. It only cuts down their population. A week later they are back. So I spray again. I have to admit, I enjoy the chirping sound at night. It sounds like you're sleeping in a tent sometimes. I even feel cruel when I exterminate these harmless creatures. However, I built the house for humans and not crickets, so one of us has to leave, and I'm the one who pays the mortgage.

Enough about the problems I encountered in the atrium area.

SNOW

As I mentioned in the beginning of this chapter, I could really only set my house facing in one direction—north-northwest. Except for the shadows in the growing area of the garden, this northern, rather than southern, exposure created only a couple of other minor problems.

One of them is drifting snow. Maryland is not known for its fierce snow storms, but we do get a couple of big snow storms every winter. This year was no exception. Since the winds blow directly onto my front door, it will be no surprise to tell you that on a couple of occasions the drifting snow packed against my front door to heights of up to 6 feet or so. Of course, an advantage to these snow storms was that the deeper the snow got, the warmer the inside became. Snow on top of the earth is a good insulation.

The problem with the snow drifting and blocking the doors was relatively easy to solve. When spring descended, I planted shrubs and trees to cause a natural barrier to block the drifting snow. In my case, as in most other underground homes, a good building location is hard to find for many reasons. Trying to find a satisfactory location with a southern exposure is like being on a treasure hunt. Most of you, however, will probably have to sacrifice that southern exposure for other benefits.

VISITORS

Don't draw the conclusion that I'm anti-social. The opposite of this is more my preference. I like to share my project with people who are sincerely interested in saving energy or who are merely curious. The only part about visitors that gets to be unpleasant is that a few, maybe 20 per cent, show up at my front door without any warning whatsoever. Not even a phone call. They just show up at the door. Well this isn't too bad since we try to keep the house presentable most of the time, and for the most part, their timing isn't too bad. If I'm really busy, the children have been very efficient about taking visitors on the "grand tour" and explaining the interesting points. However, there are a few inconsiderate people who not only show up without warning but do it at 7 o'clock on a Sunday morning. It's asking alot to be congenial under these circumstances. If you ever build an underground house, you will see that I'm not joking. The timing of some people is really off and their consideration almost nonexistent. However, I have never refused to show anyone through who has asked to see my house and answer their questions.

There is another type of person that will show up with or without an invitation. They will bring their children and dogs. If the children are well behaved, fine. But more often than not, they are renegades. These visitors are few and far between, but rest assured that if you build an underground house and it's unique to your community, you will have all types of visitors.

Most are considerate. So you see, visitors are another facet of underground living that only a year's experience could reveal.

UTILITY BILLS

Anyone investigating underground home building has checked on the utility savings—less fuel for heat, etc. Well, one year later, I can verify a few actual facts. The truth is that although I don't use a conventional heating system that uses a fossil fuel, I use only wood and a minimum of that. So I have really cut my fuel use by 100 per cent. The only negative side to the energy saving idea of an underground home is that my neighbor's electric bill is running approximately 10 per cent lower than mine. He owns a conventional normal-sized home. So I figure that I actually use only 50 to 55 per cent of the total amount of energy that I would be using if I actually had a conventional fuel bill and electric bill. This statement, however, is made with a slight reservation. Because all the traffic created by this new project has begun to subside and the constant use of power tools has slacked off, I expect my electricity consumption to drop in the coming months.

LACK OF LIGHT

Earlier in this book, I said we adapted well to the absence of windows. That, of course, was my first thought, with only limited experience living without them. I can honestly say that one year later the windows aren't even missed—not one bit. I am fully convinced that even conventional homes could do well by eliminating a few windows. With the right decoration inside, the lack of windows is a blessing. Don't let anyone try to convince you that windows are a necessary part of comfortable home living. You'll never miss them. However, there is one small catch you should be made aware of—oversleeping or losing track of time. This, believe it or not, is easy to do. Remember, once you shut a bedroom door and turn the lights out, it is totally dark. The only savior you have in the morning is the alarm clock.

During the first year, we had several occasions to sleep in one of the three bedrooms without natural light. Since it was a weekend, we didn't have to get up for work. All of a sudden the phone rang, and I grabbed it giving the party on the other end a few choice words not fit for publishing because I thought it was 4 a.m. It turns out it was nearly noon. Oversleeping happened more than once, especially to the children on school days.

PROBLEM NEIGHBORS

If by unfortunate necessity, you had to build your house close enough to neighbors so that they could see your progress, you probably had at least one neighbor give you static over your choice to go underground. The old saying, "Time heals all wounds" is true even in underground homes. You will find, just as I did, that one year later the people opposing your house have become tired of hearing themselves talk, and the neighbors who were questionable have become good friends. The official side mellows just as easily. The building inspectors will be tired of talking to you by now, especially if you haven't let them get you down.

In closing this chapter, I can safely say that the only thing rougher than the actual building construction was the first year of living inside. Things are settling down, and it is almost like living in a conventional house.

Appendix A
Weights and
Specific Gravities

WEIGHTS AND SPECIFIC GRAVITIES

Substance	Weight Lb. per Cu. Ft.	Specific Gravity
METALS, ALLOYS, ORES		
Aluminum, cast, hammered	165	2.55–2.75
Brass, cast, rolled	534	8.4–8.7
Bronze, 7.9 to 14%, Sn	509	7.4–8.9
Bronze, aluminum	481	7.7
Copper, cast, rolled	556	8.8–9.0
Copper ore, pyrites	262	4.1–4.3
Gold, cast, hammered	1205	19.25–19.3
Iron, cast, pig	450	7.2
Iron, wrought	485	7.6–7.9
Iron, spiegel-eisen	468	7.5
Iron, ferro-silicon	437	6.7–7.3
Iron ore, hematite	325	5.2
Iron ore, hematite in bank	160–180	
Iron ore, hematite loose	130–160	
Iron ore, limonite	237	3.6–4.0
Iron ore, magnetite	315	4.9–5.2
Iron slag	172	2.5–3.0
Lead	710	11.37
Lead ore, galena	465	7.3–7.6
Magnesium, alloys	112	1.74–1.83
Manganese	475	7.2–8.0
Manganese ore, pyrolusite	259	3.7–4.6
Mercury	849	13.6
Monel Metal	556	8.8–9.0
Nickel	565	8.9–9.2

Substance	Weight Lb. per Cu. Ft.	Specific Gravity
TIMBER, U.S. SEASONED		
Moisture Content by Weight:		
Seasoned timber 15 to 20%		
Green timber up to 50%		
Ash, white, red	40	0.60–0.62
Cedar, white, red	22	0.32–0.38
Chestnut	41	0.66
Cypress	30	0.48
Fir, Douglas spruce	32	0.51
Fir, eastern	25	0.40
Elm, white	45	0.72
Hemlock	29	0.42–0.52
Hickory	49	0.74–0.84
Locust	46	0.73
Maple, hard	43	0.68
Maple, white	33	0.53
Oak, chestnut	54	0.86
Oak, live	59	0.95
Oak, red, black	41	0.65
Oak, white	46	0.74
Pine, Oregon	32	0.51
Pine, red	30	0.48
Pine, white	26	0.41
Pine, yellow, long-leaf	44	0.70
Pine, yellow, short-leaf	38	0.61
Poplar	30	0.48

Material		
Platinum, cast, hammered	1330	21.1–21.5
Silver, cast, hammered	656	10.4–10.6
Steel, rolled	490	7.85
Tin, cast, hammered	459	7.2–7.5
Tin ore, cassiterite	418	6.4–7.0
Zinc, cast, rolled	440	6.9–7.2
Zinc ore, blende	253	3.9–4.2
Redwood, California	26	0.42
Spruce, white, black	27	0.04–0.40
Walnut, black	38	0.61
Walnut, white	26	0.41

VARIOUS SOLIDS

Material		
Cereals, oats ....bulk	32	
Cereals, barley ....bulk	39	
Cereals, corn, rye ....bulk	48	
Cereals, wheat ....bulk	48	
Hay and Straw ....bales	20	
Cotton, Flax, Hemp	93	1.47–1.50
Fats	58	0.90–0.97
Flour, loose	28	0.40–0.50
Flour, pressed	47	0.70–0.80
Glass, common	156	2.40–2.60
Glass, plate or crown	161	2.45–2.72
Glass, crystal	184	2.90–3.00
Leather	59	0.86–1.02
Paper	58	0.70–1.15
Potatoes, piled	42	
Rubber, caoutchouc	59	0.96–0.9.
Rubber goods	94	1.0–2.0
Salt, granulated, piled	48	
Saltpeter	67	
Starch	96	1.53
Sulphur	125	1.93–2.07
Wool	82	1.32

VARIOUS LIQUIDS

Material		
Alcohol, 100%	49	0.79
Acids, muriatic 40%	75	1.20
Acids, nitric 91%	94	1.50
Acids, sulphuric 87%	112	1.80
Lye, soda 66%	106	1.70
Oils, vegetable	58	0.91–0.94
Oils, mineral, lubricants	57	0.90–0.93
Water, 4°C. max. density	62.428	1.0
Water, 100°C.	59.830	0.9584
Water, ice	56	0.88–0.92
Water, snow, fresh fallen	8	.125
Water, sea water	64	1.02-1.03

GASES

Material		
Air. 0°C. 760 mm	.08071	1.0
Ammonia	0.478	0.5920
Carbon dioxide	.1234	1.5291
Carbon monoxide	.0781	0.9673
Gas, illuminating	.028–.036	0.35–0.45
Gas, natural	.038–.039	0.47–0.48
Hydrogen	.00559	0.0693
Nitrogen	.0784	0.9714
Oxygen	.0892	1.1056

WEIGHTS AND SPECIFIC GRAVITIES

Substance	Weight Lb. per Cu. Ft.	Weight Lb. per Cu. Ft.	Substance	Weight Lb. per Cu. Ft.	Specific Gravity
ASHLAR MASONRY			**EXCAVATIONS IN WATER**		
Granite, syenite, gneiss	165	2.3–3.0	Sand or gravel	60	
Limestone, marble	160	2.3–2.8	Sand or gravel and clay	65	
Sandstone, bluestone	140	2.1–2.4	Clay	80	
			River mud	90	
MORTAR RUBBLE MASONRY.			Soil	70	
Granite, syenite, gneiss	155	2.2–2.8	Stone riprap	65	
Limestone, marble	150	2.2–2.6			
Sandstone, bluestone	130	2.0–2.2	**MINERALS**		
			Asbestos	153	
DRY RUBBLE MASONRY			Barytes	281	
Granite, syenite, gneiss	130	1.9–2.3	Basalt	184	
Limestone, marble	125	1.9–2.1	Bauxite	159	
Sandstone, bluestone	110	1.8–1.9	Borax	109	
			Chalk	137	
BRICK MASONRY			Clay, marl	137	
Pressed brick	140	2.2–2.3	Dolomite	181	
Common brick	120	1.8–2.0	Feldspar, orthoclase	159	
Soft brick	100	1.5–.7	Gneiss, serpentine	159	
			Granite, syenite	175	
CONCRETE MASONRY			Greenstone, trap	187	
Cement, stone, sand	144	2.2–2.4	Gypsum, alabaster	159	
Cement, slag, etc.	130	1.9–2.3	Hornblende	187	
Cement, cinder, etc.	100	1.5–1.7	Limestone, marble	165	
			Magnesite	187	
			Phosphate rock, apatite	200	
			Porphyry	172	
			Pumice, natural	40	
			Quartz, flint	165	
			Sandstone, bluestone	147	
			Shale, slate	175	

VARIOUS BUILDING MATERIALS

Material		
Ashes, cinders	40–45	
Cement, portland, loose	90	
Cement, portland, set	183	2.7–3.2
Lime, gypsum, loose	53–64	
Mortar, set	103	1.4–1.9
Slags, bank slag	67–72	
Slags, bank screenings	98–117	
Slags, machine slag	96	
Slags, slag sand	49–55	

EARTH, ETC., EXCAVATED

Clay, dry	63
Clay, damp, plastic	110
Clay and gravel, dry	100
Earth, dry, loose	76
Earth, dry, packed	95
Earth, moist, loose	78
Earth, moist, packed	96
Earth, mud, flowing	108
Earth, mud, packed	115
Riprap, limestone	80–85
Riprap, sandstone	90
Riprap, shale	105
Sand, gravel, dry, loose	90–105
Sand, gravel, dry, packed	100–120
Sand, gravel, dry, wet	118–120

STONE, QUARRIED, PILED

Basalt, granite, gneiss	96
Limestone, marble, quartz	95
Sandstone	82
Shale	92
Greenstone, hornblende	107

BITUMINOUS SUBSTANCES

Asphaltum	81
Coal, anthracite	97
Coal, bituminous	84
Coal, lignite	78
Coal, peat, turf, dry	47
Coal, charcoal, pine	23
Coal, charcoal, oak	33
Coal, coke	75
Graphite	131
Paraffine	56
Petroleum	54
Petroleum, refined	50
Petroleum, benzine	46
Petroleum, gasoline	42
Pitch	69
Tar, bituminous	75

COAL AND COKE, PILED

Coal, anthracite	47–58
Coao, bituminous, lignite	40–54
Coal, peat, turf	20–26
Coal, charcoal	10–14
Coal, coke	23–32

The specific gravities of solids and liquids refer to water at 4°C., those of gases to air at 0°C, and 760 mm. pressure. The weights per cubic foot are derived from average specific gravities, except where stated that weights are for bulk, heaped or loose material, etc.

Appendix B
Underground House Statistics

- HOUSE SIZE—40′ × 90′ (3600 SQUARE FEET)

- FIFTEEN ROOMS, PLUS GARAGE

- OVER 7500 CONCRETE BLOCK USED

- OVER 250 CUBIC YARDS OF CONCRETE

- APPROXIMATELY TEN TONS OF STEEL USED

- OVER 800 TONS OF DIRT ON ROOF

- WOOD HEAT STOVE ONLY

Appendix C

Weight of Basic Materials Used in Underground Construction

Weight of Basic Materials Used In Underground Construction		
SANDY SOIL	1 CUBIC FT.	65 LB
MUD	1 CUBIC FT.	90 LB
WATER	1 CUBIC FT.	62½ LB
CONCRETE	1 CUBIC FT.	140 LB
4′ × 8′ SHEET ¼″	ACRYLIC PLASTIC	47 LB
4′ × 8′ SHEET ½″	SHEET ROCK (GYPSUM BOARD)	65 LB
4′ × 8′ SHEET ½″	PLYWOOD	55 LB
12″ × 8″ × 8″	CONCRETE BLOCK	54 LB
6″ _ 8″ × 8″	CONCRETE BLOCK	38 LB

Appendix D
Slump Test

The slump test is used to measure the consistency of the concrete. The test is made by using a SLUMP CONE; the cone is made of No. 16 gage galvanized metal with the base 8 inches in diameter, the top 4 inches in diameter, and the height 12 inches. The base and the top are open and parallel to each other and at right angles to the axis of the cone. A tamping rod ⅝ inch in diameter and 24 inches long is also needed. The tamping rod should be smooth and bullet pointed (not a piece of rebar).

Samples of concrete for test specimens should be taken at the mixer or, in the case of ready-mixed concrete, from the transportation vehicle during discharge. The sample of concrete from which test specimens are made will be representative of the entire batch. Such samples should be obtained by repeatedly passing a scoop or pail through the discharging stream of concrete, starting the sampling operation at the beginning of discharge, and repeating the operation until the entire batch is discharged. The sample being obtained should be transported to the testing site. To counteract segregation, the concrete should be mixed with a shovel until the concrete is uniform in appearance. The location in the work of the batch of concrete being sampled should be noted for future reference. In the case of paving concrete, samples may be taken from the batch immediately after depositing on the subgrade. At least five samples should be taken from different portions of the pile and these samples should be thoroughly mixed to form the test specimen.

The cone should be dampened and placed on a flat, moist nonabsorbent surface. From the sample of concrete obtained, the cone should immediately be filled in three layers, each approximately one-third the volume of the cone. In placing each scoopful of concrete the scoop should be moved around the top edge of the cone as the concrete slides from it, in order to ensure symmetrical distribution of concrete within the cone. Each layer should be RODDED IN with 25 strokes. The strokes should be distributed uniformly over the cross section of the cone and should penetrate into the underlying layer. The bottom layer should be rodded throughout its depth.

When the cone has been filed to a little more than full, strike off the excess concrete, flush with the top, with a

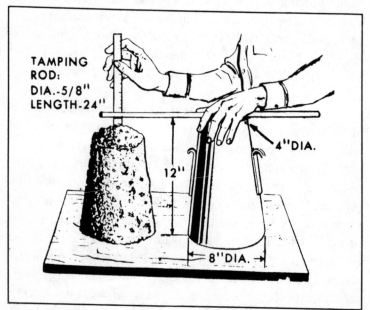

Fig. D-1. Measurement of slumps.

straightedge. The cone should be immediately removed from the concrete by raising it carefully in a vertical direction. The slump should then be measured to the center of the slump immediately by determining the difference between the height of the cone and the height at the vertical axis of the specimen as shown in Fig. D-1.

The consistency should be recorded in terms of inches of subsidence of the specimen during the test, which is called slump. Slump equals 12 inches of height after subsidence.

After the slump measurement is completed, the side of the mix should be tapped gently with the tamping rod. The behavior of the concrete under this treatment is a valuable indication of the cohesiveness, workability, and placeability of the mix. A well-proportioned workable mix will gradually slump to lower elevations and retain its original identity, while a poor mix will crumble, segregate, and fall apart.

Appendix E
Cement Statistics

Table E-1. Age Compression Strength. Relationship for Types I and III Air-Entrained Portland Cement.

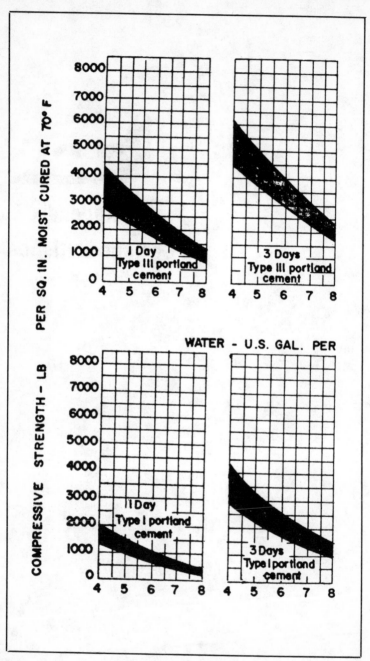

Table E-2. Suggested Trial Mixes for Non-Air-Entrained Concrete of Medium Consistency With 3- to 4-inch Slump.

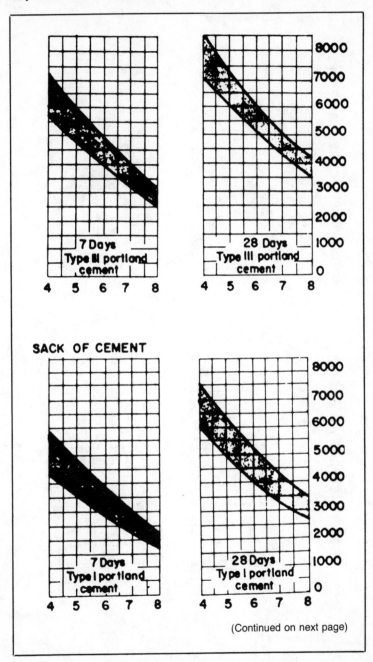

(Continued on next page)

(Continued from previous page)

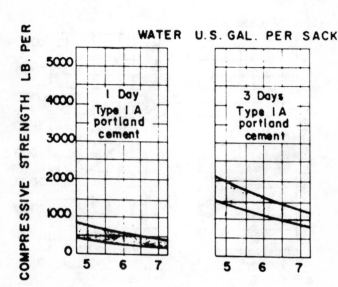

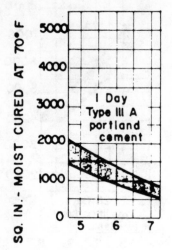

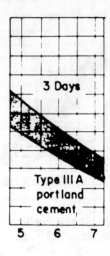

242

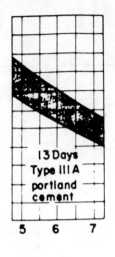

13 Days
Type III A
portland
cement

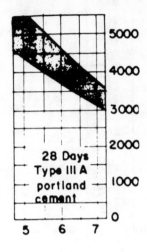

28 Days
Type III A
portland
cement

OF CEMENT

13 Days
Type I A
portland
cement

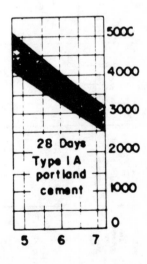

28 Days
Type I A
portland
cement

Table E-3. Suggested Trial Mixes for Air-Entrained Concrete of Medium Consistency with 3- and 4-Inch Slump.

Water-cement ratio Gal per sack	Maximum size of aggregate inches	Air content (entrapped air) per cent	Water gal per cu yd of concrete	Cement sacks per cu yd of concrete	With fine sand—fineness modulus = 2.50		
					Fine aggregate per cent of total aggregate	Fine aggregate lb per cu yd of concrete	Coarse aggregate lb per cu yd of concrete
4.5	3/8	3	46	10.3	50	1240	1260
	1/2	2.5	44	9.8	42	1100	1520
	3/4	2	41	9.1	35	960	1800
	1	1.5	39	8.7	32	910	1940
	1½	1	36	8.0	29	880	2110
5.0	3/8	3	46	9.2	51	1330	1260
	1/2	2.5	44	8.8	44	1180	1520
	3/4	2	41	8.2	37	1040	1800
	1	1.5	39	7.8	34	990	1940
	1½	1	36	7.2	31	960	2110
5.5	3/8	3	46	8.4	52	1390	1260
	1/2	2.5	44	8.0	45	1240	1520
	3/4	2	41	7.5	38	1090	1800
	1	1.5	39	7.1	35	1040	1940
	1½	1	36	6.5	32	1000	2110

6.0	3/8	3	46	7.7	53	1440	1260
	1/2	2.5	44	7.3	46	1290	1520
	3/4	2	41	6.8	39	1130	1800
	1	1.5	39	6.5	36	1080	1940
	1 1/2	1	36	6.0	33	1040	2110
6.5	3/8	3	46	7.1	54	1480	1260
	1/2	2.5	44	6.8	46	1320	1520
	3/4	2	41	6.3	39	1190	1800
	1	1.5	39	6.0	37	1120	1940
	1 1/2	1	36	5.5	34	1070	2110
7.0	3/8	3	46	6.6	55	1520	1260
	1/2	2.5	44	6.3	47	1360	1520
	3/4	2	41	5.9	40	1200	1800
	1	1.5	39	5.6	37	1150	1940
	1 1/2	1	36	5.1	34	1100	2110
7.5	3/8	3	46	6.1	55	1560	1260
	1/2	2.5	44	5.9	48	1400	1520
	3/4	2	41	5.5	41	1240	1800
	1	1.5	39	5.2	38	1190	1940
	1 1/2	1	36	4.8	35	1130	2110
8.0	3/8	3	46	5.7	56	1600	1260
	1/2	2.5	44	5.5	48	1440	1520
	3/4	2	41	5.1	42	1280	1800
	1	1.5	39	4.9	39	1220	1940
	1 1/2	1	36	4.5	35	1160	2110

(Continued on next page)

*See footnote at end of table.

Table E-3. Suggested Trial Mixes for Non-Air Entrained Concrete of Medium Consistency With 3- to 4-Inch Slump

(Continued from previous page)

Water-cement ratio Gal per sack	With average sand—fineness modulus = 2.75			With coarse sand—fineness modulus = 2.90		
	Fine aggregate percent of total aggregate	Fine aggregate lb per cu yd of concrete	Coarse aggregate lb per cu yd of concrete	Fine aggregate percent of total aggregate	Fine aggregate lb per cu yd of concrete	Coarse aggregate lb per cu yd of concrete
4.5	52	1310	1190	54	1350	1150
	45	1170	1450	47	1220	1400
	37	1030	1730	39	1080	1680
	34	980	1870	36	1020	1830
	32	960	2030	33	1000	1990
5.0	54	1400	1190	56	1440	1150
	46	1250	1450	48	1300	1400
	39	1110	1730	41	1160	1680
	36	1060	1870	38	1100	1830
	34	1040	2030	35	1080	1990
5.5	55	1460	1190	57	1500	1150
	47	1310	1450	49	1360	1400
	40	1160	1730	42	1210	1680
	37	1110	1870	39	1150	1830
	35	1080	2030	36	1120	1990
6.0	56	1510	1190	57	1550	1150
	48	1360	1450	50	1410	1400
	41	1200	1730	43	1250	1600
	38	1150	1870	39	1190	1830
	36	1120	2030	37	1160	1990

6.5	57	1550	1190	58	1590	1150
	49	1390	1450	51	1440	1400
	42	1240	1730	43	1290	1680
	39	1190	1870	40	1230	1830
	36	1150	2030	37	1190	1990
7.0	57	1590	1190	59	1630	1150
	50	1430	1450	51	1480	1400
	42	1270	1730	44	1320	1680
	39	1220	1870	41	1260	1830
	37	1180	2030	38	1220	1990
7.5	58	1630	1190	59	1670	1150
	50	1470	1450	52	1520	1400
	43	1310	1730	45	1370	1600
	40	1260	1870	42	1300	1830
	37	1210	2030	39	1250	1990
8.0	58	1670	1190	60	1710	1150
	51	1520	1450	53	1560	1400
	44	1350	1730	45	1400	1680
	41	1290	1870	42	1330	1830
	38	1250	2030	39	1280	1990

*Increase or decrease water per cubic yard by 3 per cent for each increase or decrease of 1 in. in slump, then calculate quantities by absolute volume method. For manufactured fine aggregate, increase percentage of fine aggregate by 3 and water by 17 lb. per cubic yard of concrete. For less workable concrete, as in pavements, decrease percentage of fine aggregate by 3 and water by 8 lb. per cubic yard of concrete.

Table E-4. Approximate mixing water requirements for Different Slumps and Maximum Sizes of Aggregates.

Water-cement ratio Gal per sack	Maximum size of aggregate inches	Air Content (entrapped air) per cent	Water gal per cu yd of concrete	Cement sacks per cu yd of concrete	With fine sand — fineness modulus = 2.50		
					Fine aggregate per cent of total aggregate	Fine aggregate lb per cu yd of concrete	Coarse aggregate lb per cu yd of concrete
4.5	3/8	7.5	41	9.1	50	1250	1260
	1/2	7.5	39	8.7	41	1060	1520
	3/4	6	36	8.0	35	970	1800
	1	6	34	7.8	32	900	1940
	1 1/2	5	32	7.1	29	870	2110
5.0	3/8	7.5	41	8.2	51	1330	1260
	1/2	7.5	39	7.8	43	1140	1520
	3/4	6	36	7.2	37	1040	1800
	1	6	34	6.8	33	970	1940
	1 1/2	5	32	6.4	31	930	2110
5.5	3/8	7.5	41	7.5	52	1390	1260
	1/2	7.5	39	7.1	44	1190	1520
	3/4	6	36	6.5	38	1090	1800
	1	6	34	6.2	34	1010	1940
	1 1/2	5	32	5.8	32	970	2110
6.0	3/8	7.5	41	6.8	53	1430	1260
	1/2	7.5	39	6.5	45	1230	1520
	3/4	6	36	6.0	38	1120	1800
	1	6	34	5.7	35	1040	1940
	1 1/2	5	32	5.3	32	1010	2110

6.5	3/8	7.5	41	6.3	54	1460	1260
	1/2	7.5	39	6.0	45	1260	1520
	3/4	6	36	5.5	39	1150	1800
	1	6	34	5.2	36	1080	1940
	1 1/2	5	32	4.9	33	1040	2110
7.0	3/8	7.5	41	5.9	54	1500	1260
	1/2	7.5	39	5 6	46	1300	1520
	3/4	6	36	5.1	40	1180	1800
	1	6	34	4.9	36	1100	1940
	1 1/2	5	32	4.6	33	1060	2110
7.5	3/8	7.5	41	5.5	55	1530	1260
	1/2	7.5	39	5.2	47	1330	1520
	3/4	6	36	4.8	40	1210	1800
	1	6	34	4.5	37	1140	1940
	1 1/2	5	32	4.3	34	1090	2110
8.0	3/8	7.5	41	5.1	55	1560	1260
	1/2	7.5	39	4.9	47	1360	1520
	3/4	6	36	4.5	41	1240	1800
	1	6	34	4.3	37	1160	1940
	1 1/2	5	32	4.0	34	1110	2110

*See footnote at end of table.

(Continued on next page)

Table E-4. Suggested Trial Mixes for Air-Entrained Concrete of Medium Consistency with 3- to 4-inch Slump.

(Continued from previous page)

Water-cement ratio Gal per sack	With average sand—fineness modulus = 2.75			With coarse sand—fineness modulus = 2.90		
	Fine aggregate percent of total aggregate	Fine aggregate lb per cu yd of concrete	Coarse aggregate lb per cu yd of concrete	Fine aggregate percent of total aggregate	Fine aggregate lb per cu yd of concrete	Coarse aggregate lb per cu yd of concrete
4.5	53	1320	1190	54	1360	1150
	44	1130	1450	46	1180	1400
	38	940	1730	39	1090	1680
	34	970	1870	36	1010	1830
	32	950	2030	33	990	1990
5.0	54	1400	1190	56	1440	1150
	46	1210	1450	47	1260	14000
	39	1110	1730	41	1160	1630
	36	1040	1870	37	1080	1830
	33	1010	2030	35	1050	1990
5.5	55	1460	1190	57	1500	1150
	46	1260	1450	48	1310	1400
	40	1160	1730	42	1210	1680
	37	1080	1870	38	1120	1830
	34	1050	2030	35	1090	1990
6.0	56	1500	1190	57	1540	1150
	47	1300	1450	49	1350	1400
	41	1190	1730	42	1240	1680
	37	1110	1870	39	1150	1830
	35	1090	2030	36	1130	1990

6.5	56	1530	1190	58	1570	1150
	48	1330	1450	50	1380	1400
	41	1220	1730	43	1270	1680
	38	1150	1870	39	1190	1830
	36	1120	2030	37	1160	1990
7.0	57	1570	1190	58	1610	1150
	49	1370	1450	50	1420	1400
	42	1250	1730	44	1300	1680
	38	1170	1870	40	1210	1830
	36	1140	2030	37	1180	1990
7.5	57	1600	1190	59	1640	1150
	49	1400	1450	51	1450	1400
	43	1280	1730	44	1330	1680
	39	1210	1870	41	1250	1830
	37	1170	2030	38	1210	1990
8.0	58	1630	1190	59	1670	1150
	50	1430	1450	51	1480	1400
	43	1310	1730	44	1360	1680
	40	1230	1870	41	1270	1830
	37	1190	2030	38	1230	1990

*Increase or decrease water per cubic yard by 3 per cent for each increase or decrease of 1 in. in slump, then calculate quantities by absolute volume method
For manufactured fine aggregate, increase percentage of fine aggregate by 3 and water by 17 lb. per cubic yard of concrete. For less workable concrete, as in pavements decrease percentage of fine aggregate by 3 and water by 8 lb. per cubic yard of concrete.

Table E-5. Approximate Mixing Water Requirements for Different Slumps and Maximum Sizes of Aggregates.

Maximum size of aggregate, in.	Air-entrained concrete				Non-air-entrained concrete			
	Recommended average total air content, per cent	Slump, in.			Approximate amount of entrapped air, per cent	Slump, in.		
		1 to 2	3 to 4	5 to 6**		1 to 2	3 to 4	5 to 6**
		Water, gal. per cu.yd. of concrete				Water, gal. per cu.yd. of concrete		
⅜	7.5	37	41	43	3.0	42	46	49
½	7.5	36	39	41	2.5	40	44	46
¾	6.0	33	36	38	2.0	37	41	43
1	6.0	31	34	36	1.5	36	39	41
1½	5.0	29	32	34	1.0	33	36	38
2	5.0	27	30	32	0.5	31	34	36
3	4.0	25	28	30	0.3	29	32	34
6	3.0	22	24	26	0.2	25	28	30

*Adapted from Recommended Practice for Selecting Proportions for Concrete (ACI 613–54).
**These quantities of mixing water are for use in computing cement factors for trial batches. They are maximums for reasonably well-shaped angular coarse aggregates graded within limits of accepted specifications.
†Plus or minus 1 per cent.

Index

253